Salim M. Ali

Meat
The Opium of the 21ˢᵗ Century

Author: Dr. rer. pol. Salim M. Ali, was born in India in 1954. He studied social sciences in Germany and did his doctorate in economics and social sciences. Between 1992 and 2010, he taught Sociology of Food at the University of Oldenburg in Germany. He is the author of several books on food and the worldwide eradication of hunger and malnutrition.

© Salim M. Ali, Bremen, Germany
Fist Edition October 2013
Second revised edition, April 2015
Printed in Germany

Design and layout: Suell Mües
Proofreading: Denise Beall

Acknowledgment
Thanks to Herr Joachim Harder for his support.

Cover: „Precision Farming in Minnesota", Published originally from NASA Earth Observatory under public domain, taken from Landsat 5

ISBN 978-3-73-478335-7

Books on Demand GmbH, Hamburg
www.bod.de

Dedicated to my wife Christiane

Contents

Introduction..........9
Chapter I: Eating Meat..........10
  The Value of Meat..........10
  Vitamin B12 in connection to meat consumption..........15
  Unrealistic perception of food..........17
  Carbohydrate, Protein and Calorie Budget..........19
  The alleged balanced diet..........22
  Fire, spices and salt..........24
  Slaughtering, killing, natural death, and meat and carrion eaters..........28
  Glorification of organic meat..........30
  What is lean meat?..........31
  The minced meat phenomenon..........33
  Meat as food for children..........34
  Meat in old age..........35
  Hypothesis on body parasites..........37
  Possible return of cannibalism..........38
  The Nauru syndrome..........41
  Meat for the working class..........42
  China's meat fatality..........44
  The new rich of the world..........46
  The role of the pharmaceutical industry..........47
  Health insurance may give a helping hand..........48
  Meat and the food processing industry..........49
  Restaurants..........51
  The status of man and passive food poisoning..........54
  Suffering caused by eating meat..........56
    Dental Complaints..........56
    Eye diseases..........58
    Meat and dementia..........59
    Zoonosis..........60
    Meat, virus and infectious diseases..........61
    General health problems..........62

    Insomnia and meat as a staple food..........................63
    Organ donation and meat consumption................64
  Meat consumption and the human body......................66
  Meat and gender..69
Chapter II: Meat production..71
  The theory of critical grain...71
  No association between meat consumption and world hunger...72
  The most commonly-eaten types of meat....................74
  Abundance of meat from wild game............................78
  Game meat today...79
  Meat and fodder from a global perspective.................80
  Goat breeding and desertification................................85
  Meat in the Arctic Circle...87
  Reindeer husbandry..88
  Moose..90
  Milk in nature...90
  The egg as a natural food source.................................93
  Meat and globalization..94
  USA - The World Champion in meat production......97
  Expedient producers of meat......................................99
  Meat and the fate of dogs and cats............................101
  Unavoidable contaminants in meat production........102
  Unavoidable outbreaks of animal epidemics..............104
  Separate food markets...106
  Environmental and hygiene tax on fish and meat.....108
Chapter III: The environment of meat production..........110
  A scavenger society serving the meat consumers........110
  A comparison with the life of the arable farmer........112
  Meat self-sufficiency..113
  Meat producers - the bogeymen...............................114
  The butcher and social discrimination......................115
  The glorious hunter..116
  Veterinary doctor – a childhood dream becomes a

- nightmare..118
- The patchwork of the Earth...............................120
- Animal feces and soil fertility............................121
- Meat and the world of leather products...........123
- Meat and the bottled water industry.................125
- Meat and fertile land...126
- Enmity between man and predators..................127
- The leopard phenomenon....................................129
- Alone against nature 130
- The so-called UN Climate Conference..............132
- Utilitarianism and academic negligence............134
- Chapter IV: Relinquishment of meat consumption........136
  - Meat and cultural conflicts.................................136
  - Milk-related cultural conflicts............................138
  - Meat and the world population..........................142
  - Transition to a meatless diet...............................145
  - The origin of the words 'vegetable' and 'vegetarian'..147
    - The word vegetable in English 147
    - Vegetarian..148
    - Veganism..149
  - Definitions of meat in a selection of faiths.................150
  - The different reasons for avoiding the consumption of meat..153
  - Are all non-meat eaters peaceful people? 157
  - Important incidents which can halt the consumption of meat 158
  - The avoidance of meat due to some precautions........164
  - Meat of unknown origin.....................................168
  - Professional sportsmen and the consumption of meat 169
  - Meat in hospitals...171
  - Voluntary renunciation of meat for health professionals...173
  - Meat and legal defense..174

Chapter V: Food of the future..176
    Giving up meat as a staple food......................................176
    Save existing wildlife, not dinosaurs!............................177
    Precautions for wilderness management.....................178
    Cutback in the cultivation of maize, wheat, rice and soy...180
    Recognition of other staple foods...................................181
    Sun, heat and fresh water..184
    Abuse of foodstuffs to produce conventional energy 185
    Foods which could replace meat.....................................186
    Some significant steps...189
Bibliography & References..191

Introduction

Throughout history, humans have shown an increasing preference for meat, fish, eggs and dairy products as their staple food and have even claimed this as their fundamental right. From early in the morning until late at night, animal products are served as part of a daily diet. Globalization, followed by the economic boom, has allowed the population of the world to adopt a uniform food habit based on animal products. This has resulted in the systematic destruction of man and nature. The environment with its forests, waters and wild animals has been the victim of this merciless destruction. This book examines in simple language the individual and various consequences of the growing consumption of food of animal origin.

Chapter I: Eating Meat

The Value of Meat

How valuable is meat to man? Are our bodies well adapted to ingesting nutrients from a dead animal? This question is relevant for people who eat meat as well as people who do not eat meat. Modern science can provide us with all the data and facts that are needed to answer this question.

What is meat? Meat is connective tissue muscle attached to the bone, which gives the body its shape. The body of an average animal is 76% water. The remaining 24% varies in fat and protein content, depending on the species of animal. The fat content is one third higher than the protein content. Smaller traces of carbohydrates, pollutants, cholesterol, various minerals and vitamins are also present in meat, as well as worms, nematodes, trematodes, parasites, prions, fungi, viruses and bacteria.

After water, the second biggest constituent of meat is fat. Fats are organic compounds located in an animal body in hardened or softened form. Animal fats consist principally of saturated fatty acids, which can serve as a quick source of energy. However, the effect of animal fats in the human body is very controversial and judged to be predominantly negative. Animal fats are not regarded as being healthy or having a neutral effect, but rather as being harmful.

Quantitatively, the third largest constituent of meat is protein. Proteins are the building blocks of the body and they consist of amino acids. Amino acids are compounds made up of carbon, hydrogen, oxygen and nitrogen. Proteins are found not only in meat, but in all foods. Without proteins the cells of organic products could not be produced. Protein deficiency triggers diseases such as kwashiorkor or marasmus. These diseases are typical 'hunger diseases' that are caused by chronic undernourishment and malnutrition.

Meals that are rich in protein produce stools with abundant protein. The dry matter of the feces of well-nourished people is made up of up to 30% crude protein. Droppings of farm animals with different feeding behavior yield different levels of protein. Cattle eat mostly low-protein food such as grass or hay. Therefore cow dung contains only 8% crude proteins. In contrast, the feces of pigs and chickens contain 24% to 30% crude proteins (AFRIS/Feedipedia). The water in meat leaves the body as urine, sweat and in feces. The protein content leaves the body in the stool, and only the fat content, which is considered unhealthy, remains in the abdomen.

Proponents of meat in a normal diet, regard meat as a significant source of protein and they would argue that a deficiency of meat is harmful to health. To counter this, the question can be raised why members of the Jain faith, who have not consumed meat for thousands of years, do not suffer from protein deficiency?

Apart from fat and protein, there are few other significant nutrients in meat. The carbohydrate content is extremely low, less than one percent, and can therefore be described as negligible. Depending on environmental conditions, the level of pollutants present in meat may vary. Harmful pollutants such as lead, cadmium, cesium isotope 137, dioxins, nitrates, nitrosamines, polychlorinated biphenyls and mercury are often present in meat. Growth hormones and antibiotics against diseases in animal feed can be transmitted through the food chain to humans.

Meat contains cholesterol. An increase in cholesterol levels in the blood can lead to cardiovascular disorders. Meat also contains minerals, mainly sodium, calcium, iron and phosphorus. The iron content is of importance. Like meat, vegetables also contain iron, but the iron content in meat is heme-iron which can be absorbed by the body more quickly and effectively than iron from plant foods.

Iron is an essential nutrient for plants, and taken with other nutrients as a compound, it is a divalent iron. Ingested through the food chain, divalent iron is stored in animal blood as ferric iron. This trivalent iron in flesh and blood is absorbed by the human body more quickly and easily than divalent iron from plant foods. For this reason, many nutrition experts including physicians recommend eating a lot of meat, thereby avoiding the risk of iron deficiency symptoms such as severe fatigue. This view throws up questions with no ready answers. If plant food has an inferior iron content, how can it be that the meat from herbivores is rich in iron? From where do herbivores get the high iron content in their bodies? Or, if iron deficiency causes fatigue, how can a lion, a pure carnivore, sleep 18 hours a day? Moreover, is it essential to look for iron in meat, or is iron also present in plant foods? Does the flesh of a carnivore have a higher iron content than that of a herbivore? All of these questions are ignored by proponents of meat-eating.

Numerous studies have shown an increased iron content in the body to be far more harmful than the effects of iron deficiency, and can result in an increased risk of diseases such as cancer, diabetes, cardiovascular disorders and hemochromatosis, which destroy organs such as the liver, heart and pancreas. (Federal Institute for Risk Assessment, No. 016, Berlin, 2009).

The most important vitamin in meat is vitamin B_{12}, also called cobalamin. However, micro-organisms present in the digestive tract produce vitamin B_{12} and deliver the required amount to the body as a function of symbiosis. Despite this, it is claimed that the consumption of meat supports the cobalamin balance in the body. It is interesting to note that the members of the Jain faith, who consume no meat, do not suffer from cobalamin deficiency. Other vitamins which exist in meat, such as retinol and ascorbic acid, can be found much more in plant food than in the flesh of animals or in animal organs such as the heart, liver or kidney.

The surface of an animal's body as well as its digestive tract are home to a number of worms. Nematodes are roundworms and trematodes are flukes. In favorable conditions, they can spread to almost all parts of the animal's body. Fleas, lice, mites and ticks are among the larger parasites feeding on the surface of an animal's body. Prions are toxic and dangerous proteins that can cause illnesses such as Creutzfeldt-Jakob disease. Furthermore, there are micro-parasites such as bacteria, fungi and viruses in an animal's body. All these parasites can be transmitted to humans through meat and cause diseases.

The valuable nutrients in an animal benefit the animal itself and are not meant for humans who first have to kill the animal, then dismantle and cook it before eating it. The nutrients in an animal may not be transferable to humans through digestion. Every individual or animal absorbs nutrients from a long-term intake of food or they produce these nutrients themselves. Whether the valuable molecular compounds that are present in an animal fulfill the same needs in humans is unclear. It cannot be assumed that animals, who can meet their needs thorough their own metabolisms, are necessarily useful to man.

At a time when man practised cannibalism, he could justify this act in two ways. Firstly, killing and then eating the enemy acted as a demonstration of his power. Secondly, through the consumption of a dead body by family members, the deceased was taken in their own bodies and the dead person continued to live inside them. Because such primitive peoples had no idea of heaven, this was a way of establishing eternal life after death. In both these instances of cannibalism, food was the principal issue. The idea of living on in another body is based on superstition, but science supports this thesis because of the valuable nutrients which are thus absorbed by the living body. In these societies, it was the wish of the deceased to be eaten by their family members. But no animal wants to be eaten by a human, and the fear and resentment of an animal at being eaten by a human causes biological changes at the molecular level. Because the human body is not well suited to the consumption of meat, these kinds of 'frightened' amino acids may cause ailments.

In the past, the traditional way of acquiring meat for consumption required a lot of effort. Hunting, and slaughtering were extremely difficult and demanded a lot of labor. The physical strain of hunting an animal justified the consumption of its flesh, even in large quantities. However, in modern life, meat is supplied to consumers as a finished product, similar to the biblical manna from the heaven, and not as an animal in the wild or in a shed. The rage aroused in the animal through its slaughtering is not apparent to the consumer. However, this type of consumption is usually found in prosperous countries and this relatively effortless and peaceful method of consuming meat results in nutritional diseases.

What are the reasons which motivate man to eat meat? The smell of a burning animal in a forest fire or the smell of a dead animal in nature awaken those instincts which create the desire to devour meat. Up to the point where rigor mortis sets in, meat is odorless. From the beginning of the decaying process, meat starts to smell. It becomes more intense and ends with complete decomposition. Meat also gives off a smell when it being heated and the addition of fat, salt and spices makes the consumption of meat more desirable.

Why has man selected eating meat as a means of acquiring valuable nutrients? In order to obtain certain vitamins and minerals the following long detour was chosen: first, the forests were cleared, predators were killed, pests and vermin were destroyed, fodder was planted, watered and fertilized, animals were bred, butchered, cooked and then eaten in order to ensure the intake of certain vitamins and minerals. For example, a plant eaten as fodder provides iron for a cow, and the cow, once slaughtered and eaten supplies iron to humans - a complicated procedure to gain nutrients.

What is so important in meat that man absolutely has to eat it? Is meat a natural food or is it the carcass of a dead or slaughtered animal which has to be buried or disposed of? This disposal may take place in nature, through natural waste disposal or through our stomachs. Now to the question: Which method of disposal is better? If enough other foods are available, why should we dispose of carcasses of dead animals by means of our sensitive digestive system?

Vitamin B12 in connection to meat consumption

Anaemia is a common disease of mankind. Excessive amount of blood loss due to injuries was an additional problem for physicians. George Whipple carried out an experiment in the year 1920 with many healthy dogs. He cut their veins so that the dogs lost a quantitative amount of blood. Afterwards he fed them with a number of feeds of plant and animal origin. Some of these feeds such as liver, kidney, meat and apricot showed a strong stimulating effect. Finally he chose liver, fed the experimental dogs with this particular feed and noticed a faster recovery.

George Minot and William Murphy copied the idea of Whipple and carried out experiments on patients suffering from pernicious anemia. Minot and Murphy published a report in 1926 about the treatment of pernicious anemia by means of a special diet consisting of liver, kidney, meat and vegetables, but the latter two had to be taken in larger portions. Subsequently they propagated a liver diet as the only way to win the battle against the pernicious anemia.

A sickening experiment began. Anemia patients had to eat only liver several times a day; even raw liver as patients diet was no exception. The experiments were carried out in a number of hospitals in the United States and eventually it was adopted worldwide as the liver diet or liver therapy against pernicious anemia. The consumption of liver, kidney and meat has been proposed as the prime means of blood formation in the body.

Based on this discovery, the medical team of Whipple, Minot and Murphy was awarded with the Nobel Prize for Medicine in 1934 (Professor I. Holmgren, Presentation Speech, The Nobel Prize in Medicine 1934). Up to this time there were no vegans in the Western world, to whom anemia could be labeled. There were some vegetarian organizations, but they were not affected by the charge of meatless diet in connection to anemia.

There have long been speculations about the contents present in liver or meat, which supports the formation of red blood cells. Using X-ray diffraction, the English chemist Dorothy Crowfoot Hodgkin managed to isolate the molecule called B12 in 1956. Thereupon in 1964 she was awarded the Nobel Prize in Chemistry (Professor G. Hägg, Presentation Speech, The Nobel Prize in Chemistry 1964).

It is claimed that the B12 molecule ($C_{62}H_{88}CoN_{13}O_{14}P$) is too large to penetrate the human intestinal wall. This claim seems to be quite dubious, because the human intestinal wall is not more tough than the intestinal wall of other mammals. In fact, the Cobalamin supply can be disturbed by various causes, such as by excessive alcohol consumption, strong medications, illness or malnutrition. As a result the microorganisms may not be able to produce sufficient cobalamin.

The excessive meat eating idea of Whipple, Minot and Murphy of the 1920s was brought back, so that the vitamin B12 budget remains intact. But there is no mention that the human intestinal wall is hard-permeable for Cobalamin molecule of meat origin. With the help of microorganisms vitamin B12 can be produced artificially at a very reasonable price, which can also be taken as a meat substitute. There is a booming market of vitamin B12 products. These are not drugs, but vegan and vegetarian food products such as jam, soy milk, fruit juice or granola bars. A healthy intestinal flora is more important than the fears about vitamin B12 deficiency.

Unrealistic perception of food

What is the value of nutrients? Science conveys an abstract, new concept of nutrition, which is difficult for ordinary consumers to understand. The market-based nutritionist travels through a galaxy of molecular compounds to arrive at a definition of our daily food. They know the modality of even the smallest food particle. For example, they can reveal to us that the essential vitamin B_{12} can be found in the centre of a calf's liver (58.2 micrograms per 100 g of calf liver). Male calves are slaughtered shortly after birth because they are not economically viable and not because they are a source of vitamin B_{12}. Despite this knowledge, many people, including food experts, suffer from modern nutritional diseases.

Terms such as carbohydrates, proteins, calories, saturated fatty acids, unsaturated fatty acids, or amino acids are invisible, imperceptible and unimaginable things, which people do not realize make up their daily food. All products in the world, whether organic or inorganic, are composed of numerous components. The modality of the ingredients results in a new or different product. For example, over 95% of a cucumber is water, and still a cucumber is not mistaken for water. If a person eats only cucumbers, he may well survive for a longer period of time. But if the same person were fed with less than 5% of the chemical ingredients of cucumber such as carbohydrates, proteins, unsaturated fatty acids, fiber and over 95% sterile water, he would not survive because his digestive system would be disturbed and his metabolism would be hampered. If a plant were fed in the same way, it would survive for a long time. But if the plant were supplied with unprocessed solid cucumbers, the plant would die in a few days. It can be understood from this simple and practical example that food for plants is made up of invisible molecular compounds, whereas food for humans and animals are visible bulk products. Equating plant food with human food is a regrettable scientific mistake.

It is the digestive system that determines which nutrients are absorbed by the body and which must leave the body as feces. Choice, taste, and selection of foodstuffs are factors involved in food preparation and the digestion of food completes the process of food intake. The more crucial part lies in the selection of which foods are to be eaten. Knowledge of the nutritional content at a molecular level is an asset, but it does not mean that a person with this knowledge can influence the metabolism of the digestive tract. A person is responsible for ingesting appropriate and healthy food and not for doing the work of their metabolism.

Invisible nutrients as food for plants: Plants do not have a digestive tract or a central nervous system. The most significant difference between plants and animals, is that animals consume food in bulk, which undergoes the digestive process and any unused food is excreted. In contrast, plants do not consume food in bulk, but rather, molecular compounds are collected through roots and produced by leaves. The water which plants require, is used primarily for the transportation of nutrients. The roots take up water with nutrients and distribute this to all parts of the plant. A large part of the water evaporates through its pores.

Plants absorb nutrients as molecular compounds and therefore do not need a digestive tract. However, in humans and animals, molecular compounds are a product of their metabolisms, and this results in the production of excrement. Unfortunately, modern nutritionists have suggested the same food needed by plants can serve as healthy food for humans. Compulsory information on food labels lists valuable nutrients such as carbohydrates, proteins, vitamins, minerals or even unhealthy nutrients such as fat or sugar. It is useful to know because one should not take in a lot of sugary or fatty foods, but how can a person know what he is ingesting at a molecular level? Will these valuable nutrients remain in the body to perform their appropriate task, or will they leave the body through the disposal routes as quickly as possible?

Food for an unborn baby: The fertilized ovum develops into an embryo and then into a foetus. Simultaneously the placenta develops to provide the baby with nutrients. Although the foetus developes a digestive tract, no digestion takes place because the foetus does not receive bulky food. The placenta takes over this role and up to the moment of birth it supplies the baby with nutrients, like the roots and leaves that feed a plant. The umbilical cord is not a hose for transporting food, but the nutrients are transfered to the foetus via diffusion. These mechanisms cannot be influenced from outside, nor can extra nutrients such as water, oxygen, or vitamins be added. The only way to maintain the nutritional function of the placenta is through the mother's usual intake of food. Here it becomes clear that it is impossible to feed a foetus directly with molecular compounds.

Carbohydrate, Protein and Calorie Budget

Carbohydrates: Carbohydrates consist of three organic elements: carbon (C), hydrogen (H) and oxygen (O). Different molecular compounds represent different carbohydrates. These three elements make up the majority of the plant world. However, the carbohydrates in nature are not visible because they combine with other substances to produce a leaf, a root or a fruit. In general, carbohydrates are soluble in water and taste sweet. Our staple wheat consists of 60% carbohydrates and rice 78%. However, rice or wheat in thier raw state are not water soluble nor do they have a sweet taste. The molecular separation takes place in the digestive tract. It does not matter how much rice or bread a person consumes, the body will only absorb as much as it needs. The rest will be excreted. It is misleading to refer to bread as a carbohydrate.

Protein: The consumption of protein-rich foods is at the forefront of our diet. Fish, meat, eggs, legumes and dairy products are important sources of protein in the modern diet. Proteins consist of amino acids. Amino acids are structural elements of carbon (C), hydrogen (H), oxygen (O) and nitrogen (N). A combination of different atomic numerals results in different amino acids. For example, isoleucine ($C_6H_{13}NO_2$) is an important amino acid for our bodies whereas histidine ($C_6H_9N_3O_2$) is less important. We cannot influence the formation of these molecular compounds, which we do not see, feel or can affect in any way. No matter what type of protein or how much protein food we eat, the body retains only what is needed. The excreta of humans and pigs for example contain 25% crude protein, and chicken manure in its dry state even contains 30% crude proteins (FAO / AFRIS, Useful references: 25, 184, Manure, Rome 2012). If proteins are as valuable as it is claimed, why does the body throw out such a large amount of protein as excrement?

Calorie budget: Energy in food is calculated as a caloric unit. A calorie is the energy required to heat one gram of water from 14.5° C to 15.5° C. In the human body this calculation is based on the quality of food rated as a kilocalorie (1 kcal = 1,000 cal). For example, 1 kg of water without additives has 0 kcal because water is a neutral substance such as air and has no nutrients. On the other hand, 1 kg of Cola has about 440 kcal because Cola contains sugar. Another example: 1 kg of potatoes contains 790 kcal, 1kg of veal 1200 kcal. 1kg of potatoes and 1kg of veal has 780 grams and 760 grams of water respectively. But it is not clear to what extent the two different products supply energy to the body, and how it is calculated. It is not clear if the body temperature is different from one food to another or if one particular food lasts longer in the body than another. Other factors such as body weight, age, mobility, stability, room temperature or different health conditions must also be taken into account. Numerous devices and measuring instruments are available to measure and calculate other units, but not food input. Despite a passion for counting calories in our affluent society, obesity remain as widespread health issue.

The assessment of food in calories is not very helpful, because the energy received by a person from a particular food differs, depending on age, body size and health. Every person processes every food differently. Taking in food with a higher kilocalorie value does not necessarily mean consuming more energy. It just means that one has taken in more calories. Since the feces also have a certain calorie value, it is not clear how much energy is actually absorbed from the food consumed. This is an area which needs investigation.

Calculating calorie budgets was invented in 1824, proteins were discovered in 1838 and carbohydrates were detected in 1844. These two century-old findings constitute our basic information about food. Profit-oriented science can continue to define food nutrients, but an apple or a potato will continue to be simple foodstuffs for consumers.

Foodstuff or medicine: Medicine is not regarded as food nor are its ingredients referred to as nutrients. Medicines are drugs that also consist of proteins and minerals. The famous antibiotic penicillin ($C_9H_{11}N_2O_4S$) is a compound of different elements with different variables. Here the concept of active nutrients differs from medicinal ingredients, because nutrients provide energy to the body and drugs do not. Very strong spices are identical to drugs, in that an overdose can be harmful. The smallest inclusion of drugs must be listed, defined and disclosed, because these substances may perform changes in a body. Almost all drugs are toxic substances and so cannot be regarded as food. This raises the question why the information on food labels are defined in a similar way as drugs, although food never occurs in this state.

Neither the food industry nor cooks use separate components such as proteins, trace elements, vitamins, carbohydrates, fiber or individual fatty acids to produce meals. They use commodities such as grain, oil, or potatoes but food labels describe the final product at a molecular level. This practice causes not only confusion, but it is also an attempt to seize economic power in the area of food production. The food industry will certainly play an identical role in the foreseeable future to the pharmaceutical industry and it may be that independent private food production will be declared illegal.

The alleged balanced diet

A precise definition of a balanced diet for the average consumer does not exist. It is the pure invention of an affluent society, where a variety of food in unlimited quantities is available. Depending on availability, the list of ingredients making up a balanced diet can be extended. For example, producers of nuts (cashew, walnut, pistachio, almond, peanut etc) recommend consumers to eat a handful of their product on a daily basis. But the human body is unique so that a basic rule for food intake is a purely theoretical concept. An empirical survey on what constitutes a balanced diet is difficult to carry out because a number of criteria such as age, sex, individual body condition, the effect of the food, environment and so on must be taken into account. This idea of what constitutes a balanced diet leads to confusion with no useful results. Perhaps five guinea pigs from the same litter would be helpful for such a study but in humans, such a survey would be hard to carry out.

An unbalanced diet is caused by two main factors. Firstly, it may be due to poverty or a scarcity of food. Secondly, it may result from a preference for some particular food. The former is outside of human influence, whereas the latter may occur for a variety of reasons. For example, many people eat only white bread or prefer a diet of pure meat. Perhaps a person can only stomach white bread and tolerates no other food and the person is grateful that white bread is available. The example of the pure meat eater should be examined carefully. In what environment does he live, for example, above the Arctic Circle, in a desert or in a modern city?

The giant panda can grow to be 180 cm tall, weigh up to 160 kg, can live up to 30 years and feeds exclusively on bamboo leaves. The giant panda survives on a single food source because for centuries his only natural habitat has been bamboo forests. He does not suffer from nutritional deficiencies caused by an unvaried diet. This should give believers in a 'balanced diet' food for thought.

Donald Watson (02.09.1910 - 16.11.2005), the founder of the Vegan Society, consumed no meat or animal products from the age of 14. He was never sick, did not require medical attention, did not wear glasses, never took any medicine, led a decent professional life and reached the age of 95. In the last years of his life, he hiked in the high mountains in the north west of England. Examples abound of such individuals: inhabitants of certain islands, members of some religious communities, desert dwellers or hill tribes have achieved longevity with very limited food.

A person's daily diet was not always a purely individual choice, but was also an environmentally-dependent development that took place across vast stretches of time. People ate what grew around them during their life time. The modern industrial age has brought this traditional diet principle into utter confusion with food products imported from outside regional areas. For example, fruits from New Zealand are eaten regularly in the northern hemisphere. The diet of modern society is filled with foods of animal origin and successfully-grown regional plant food is often dismissed as tasteless.

A list of foods which make up a balanced diet consists mainly of cereals, vegetables, fruits, milk products, eggs, fish and meat. It is emphasized that one should eat fresh vegetables, fruits, cereals and as a side dish various animal products. This is an attempt to justify the consumption of meat, fish and milk. How can meat be considered part of a balanced diet, if man is not a predator and cannot eat meat in its natural state? This also applies to fish. As man cannot catch fish without tools and cannot eat fish without first filleting it, how can fish belong to a balanced diet? Furthermore, milk, which is produced to nourish babies of the same mammal species, is considered to be an essential part of a balanced diet.

A theoretical list of foods which make up a balanced diet is not global, but depends on the resources available and their natural compatibility with the human body. The list of foods essential for a so-called balanced diet, in which animal products feature largely, is at the same time also a list of the causes of modern nutritional diseases.

Fire, spices and salt

Under extreme climatic conditions, such as in hot or cold deserts, where there is little or no vegetation, people had to eat meat in order to survive. People would certainly have preferred plant food, but they had to eat meat instead. In the colder areas, where no fuel was available, people ate raw meat and in the hot deserts, where there was also no fuel available, one could dry meat in the sun, but in both cases the consumption of meat was very difficult and the supplies were very limited. For these reasons alone, the number of people who ate raw or dried meat was very small, as small as the number of predators. If this had not been the case, the desert and polar populations would have increased substantially, the wildlife would have become decimated and the resulting famine would have put an end to their existence.

Fire: We do not know why human beings became separated from the animal world. However, there are two activities that man can do and that animals cannot. These are using tools and the knowledge of how to make fire. Some animal species use some tools such as sticks to move something, hard ground to break something, a large leaf to protect themselves from the rain, or materials to build nests. But no animal has the ability to make fire. In the film adaptation of the "The Jungle Book", a children's story written by the Nobel laureate Rudyard Kipling, the monkeys ask Mogli, the jungle child, how human beings make fire. With the legacy of making fire, man was expelled from the animal world.

If monkeys, the alleged relatives of humans, had been able to make fire, they might have followed the same development as man. With fire, man became self-sufficient as regards food, and everything that could not be directly consumed, was treated with fire. There are thousand- year- old fire trails in the wilderness of Australia, that were created for the purpose of hunting. People started eating animal carcasses following a natural forest fire. Later they torched the wilderness to create an indirect form of hunting. Common predators do not eat burned or charred carcasses. Only man adopted this practice and the modern barbecue culture is a genetically inherited practice.

Heat treatment is the basic method of meat preparation for humans. Fire makes meat more tender, reduces the risk of infection, reduces the water content of meat and finally renders it tastier than in its raw state. Why people in polar regions do not get ill from eating raw meat, is certainly because of the lower temperatures, which reduce bacterial decomposition. People who live in warmer climates, believe that treating meat with heat, destroys all the pollutants and pests and secures the most valuable ingredients that are available in meat.

Spices: Spices are pungent ingredients used in a small quantities. Herbs, roots, fruits, flowers, seeds, or minerals can be used as spices. Many of these spices are poisonous if consumed in large quantities, so only very small doses can be used. They impart a distinctive flavor and aroma to food when cooked or eaten. In this way inedible food is made palatable enough to eat.

The meat that was eaten originally by man was venison and the meat smelled different depending on the species of animal it came from and what the animal had fed on. Venison has an odor of urine and feces, tastes watery, moldy and is tough.

Spices helped to transform the meat into something edible. Hot spices are used to make a person swallow tasteless food more quickly and fragrant spices are used to hide the unpleasant smell of meat. A mixture of spices was developed for use with different types of meat and so spices became a constant companion of meat and fish as food for human consumption. Stronger-tasting spices such as garlic, onion, chilli, pepper, cardamom, cloves, ginger, cinnamon, cumin, saffron or turmeric are mainly meat spices. A mixture of these spices added to meat and subsequently treated with heat impart an appetizing aroma and flavor to meat. The reason why non-meat eaters also consume these spices is because their ancestors were meat eaters and they have taken over the same spices to flavor vegetarian dishes.

Salt: Salt used to be one of the rarities that human beings truly appreciated. Until the beginning of industrialization, salt was termed "white gold". It was too expensive to be used to preserve meat or fish. Only in special circumstances, such as going on a long journey, was meat preserved with salt. The common medium of exchange for gold was salt and in many parts of the world gold and salt trades were directly linked in their development. Famous cities, routes and stories emerged from the salt trade. In many countries a tax on salt provided governments with a significant revenue and up to the 20th century, numerous salt wars were fought.

Industrial salt production allowed easier access to salt and this landmark trend has transformed salt into one of the cheapest commodities in the world. The preservation of fish and meat with salt is a very recent development and was not practiced in the past. The consumption of caviar as a delicacy is a new fashion. Caviar was originally valued only because of its salt content. A small spherical grain of caviar was placed on the tip of the tongue because of its salty taste. In reality, caviar is not a delicacy, but rotting fish eggs that taste like salt and smell of rotten fish. Why were smelly fish eggs chosen to become a delicacy and not another, more fragrant product? The reason was that salt was very expensive and one should not consume a lot of it.

Industrially-produced salt enabled a new type of meat consumption. Meat that was previously swallowed at high speed, began to be chewed in the mouth slowly and pleasurably. This new taste of meat inside the mouth is not the taste of the meat itself, but the taste of salt and the different types of meat which lead to a psychological appreciation of meat. Salt has become an inseparable companion of fish and meat. Virtually all fish and meat products are prepared with enormous quantities of salt and people consume more salt than ever before in human history. Salt added to rotten fish and meat has become a new delicacy. Salt impedes bacterial decomposition and so it has been used in varying amounts for the purpose of preserving fish and meat. Thereafter, heat treatment gives the fish and meat a special salty flavour and smell, which people find appetizing.

Carnivores do not need any spices, salt, or heat treatment in order to consume meat. They hold the meat with their claws, tear it with their incisor teeth and devour it without chewing. A crucial difference between carnivores and herbivores is that herbivores love salt and search for it in the wilderness. In contrast, predators show no inclination for salt. The suggestion that carnivores get salt through eating herbivores is absolutely wrong, because the flesh of neither carnivores nor herbivores contains table salt, nor is one saltier than the other. The most common mineral content of the flesh of land-living mammals are potassium, phosphorus, sodium, calcium and iron, and these substances do not make sodium chloride, in other words, table salt. If herbivores consumed salt, this would not be stored in the muscles or anywhere else in its original form as a chemical compound, but it would be converted by the metabolic process. Without the addition of salt as an ingredient, meat-eating would be less widespread.

Slaughtering, killing, natural death, and meat and carrion eaters

Because man is not built anatomically for hunting; he was more of a scavenger than a natural hunter. But he was not a carrion eater, like hyenas and vultures, whose anatomies are suited to this type of feeding. Man is more like a crow, raven or pig, which cannot process a large carcass. As long as man did not use any tools, his consumption of dead animals was more arduous. The use of tools enabled the hunting, killing and slaughter of livestock. The only difference between killing and slaughtering is that when an animal is slaughtered, it is under human supervision. This is not the case in a killing. Animals that run away when chased are killed from a certain distance. Animals that can be held by man are slaughtered. The term "slaughter" is derived from religious ritual. There are rules for slaughtering in faith communities where meat is approved as a source of food. Animals are slaughtered according to religious rules, either as an offering or simply as food. Without religious rules, there is no difference between killing and slaughter. For this reason modern industrial slaughtering can be termed killing rather than slaughtering. However, in both cases an animal dies and it makes no difference to the carcass whether the animal has been killed or slaughtered.

Natural death is not the same as killing or slaughter. Natural death can have an age-related cause or be due to illness. But in both cases, rigor mortis sets in after the same interval of time. After death, a chemical process begins within few hours, which gradually makes the carcass rigid and inflexible. In this condition the flesh is very tough and not very palatable. But, depending on the temperature, the decomposition process starts between 24 and 48 hours afterwards and the flesh begins to soften. This decomposition process is called by butchers ripening or maturing. In fact the meat is decomposing because the bacteria begin in a natural way to decompose a dead animal. The quality of the meat and its constituents such as water content, fat or proteins remain the same in all states of death. The meat of an animal which has died in a natural way does not have less protein than a slaughtered animal. What makes the meat of a slaughtered animal better than that of an animal that has died naturally? In the plant world, ripe fruit falling from a tree usually tastes better than fruit which has been picked. A predator usually hunts an old herbivore because older animals taste better than the younger ones. Scavengers are also to some extent predators, but they prefer animals that have died of old age, because, like ripe fruit, they have a stronger taste. It would be better for meat consumers to eat animals that have died of old age rather than young animals that have been killed.

Meat comes from a carcass and processed meat bears no indication whether the animal was killed, slaughtered or died in a natural way. There is a strong link between predators and scavengers because predators can also be considered scavengers when they are fed in a zoo or in a circus cage. In nature, lions or tigers do not usually eat dead animals that they have not killed themselves, but when such dead animals are cut up and thrown to lions or tigers, they eat them. So, to the predator it makes no difference whether the animal was killed for meat consumption or simply died. Many animals classified as scavengers, like hyenas and vultures, are occasionally predators and almost all predators such as wolves, lions or bears are also scavengers.

Terms such as carrion or meat have been invented by man. In nature there are no such differences. If a lion chases an antelope, kills it and eats it, he is regarded as a carnivore, and those waiting opportunists that eat the same meat are considered to be scavengers. Predators that can move faster prefer living animals because the prey moves. A cheetah is not classified as a scavenger, because it devours its prey as soon as it has been caught. This view is wrong, because the cheetah is the fastest predator in the world, and so it is easier for it to chase its prey. Their habit of eating quickly is certainly the reason why cheetahs can ingest their prey before rigor mortis sets in. Highly responsive animals usually eat faster. This has nothing to do with the fear of loss of prey or the conversion of fresh meat to carrion. The cheetah is a scavenger in captivity, whether it is fed in a zoo or in a circus cage with cadavers.

When man lived in the animal world, he was dependent on predators or scavengers to steal a piece of partly processed meat. Civilized man possesses tools for processing meat. In both situations, the flesh of an animal is still a carcass and not a piece of cultural heritage which is called meat. Regardless of his situation, man remains as scavenger or carrion eater. Certainly many people will not eat meat if meat is perceived as parts of the body of a dead animal.

Glorification of organic meat

Organic meat is another reason given for eating without risking injury to health. In addition, the superior taste and quality of organic meat is emphasized. A lot of things have been developed for the welfare of organically-bred animals. Animal welfare is the new legitimacy to justify animal husbandry. According to natural law, humans are not entitled to breed animals and for this reason alone, animal husbandry can never be a fair undertaking. In so-called species-appropriate husbandry, animals are fed with clean and organically-grown fodder, and allowed plenty of space to live and move. Calves are fed with milk at the beginning, and on some organic farms they may even suck milk from their mothers. Often piglets may suckle milk from the sow as long as they want and organic poultry are allowed to roam freely under the blue sky. Life could not be more comfortable for these lucky animals. However, time passes, the animals must put on weight and then the animal should preferably walk to the butcher, rather than be transported there in energy-consuming vehicles. This is called producing organic meat.

The grateful animals enjoy a magnificent standard of living in their short lives. On conventional chicken farms, animals usually live only for 35 days before they are delivered to the slaughterhouse. In organic poultry keeping, chickens may live to be 53 days old, before they are slaughtered. Mammals such as pigs, sheep and bulls are castrated relatively painlessly. Without castration the odor of urine would spoil the taste of the meat. Pigs and calves on organic farms are allowed to live slightly longer than in conventional farming. When not bred for food consumption, a chicken can usually live to be 9 years old. Pigs, cattle and sheep reach an average age of 24. Even in the organic meat industry, the animals are allowed to live less than 2% of their natural life span. So, organic meat is pure baby meat and the young animals are even neutered.

A tribe called the Bhogi, which means a person who savours life, lived in South Asia up to the late Middle Ages. The Bhogis were very much respected in society. They were loved by all, no one become angry with them, they were provided with the best food available, they enjoyed great freedom among fellow men and women and they did not have to work for their livelihood. The only thing required of them was that one of the Bhogis had to be sacrificed every year as an offering to the gods (Gait, Edward; A History of Assam, p. 59, Calcutta, 1933). In the same way, organically-bred animals are well-treated during their relatively long lives, but all end up in a slaughterhouse.

Organic meat can be naturally organic, if an animal lives in freedom, becomes old without mutilation and dies of old age rather than being slaughtered. Only then can the meat be regarded as matured organic meat.

What is lean meat?

Lean meat is considered to be low-fat, high-protein food that promotes good health and keeps the body fit and slim. Lean meat is found in every animal, but skill is needed to remove it. If meat consumption is rated as harmful, lean meat proves the opposite. In recent decades the demand for lean meat has increased dramatically and the global demand for lean meat has no limit.

The demand for lean meat is directly related to the present economic growth. The greater the purchasing power, the higher the demand for lean meat because it is more expensive than normal meat with fat. It does not matter to the consumer what methods are applied to produce lean meat. The main issue is that it is good for their health. Another new trend in lean meat consumption is organic lean meat, where the conditions are even more complicated than those for the production of conventional lean meat.

Lean meat is the muscle tissues of an animal with very little fat. The quality of lean meat is typically rated at having a 10%, 20% or 30% fat content. Biologically it is not possible to generate an absolute fat-free meat. To fatten an animal, it must be fed regularly either with organic feed or conventional feed. The feed is converted into three main substances - water, fat and protein. Faster growth results in an increase in the fat and water content, and less protein. Omnivores such as pigs and poultry excrete large amounts of food proteins as excrement. Herbivores such as cattle or sheep consume low protein-containing feed and excrete accordingly lower amounts of protein. This does not leave much protein in the animal body which has to be added by feed input. The average protein value in pork meat is 13% and in chicken and beef it is 19%. In fact, lean meat is water meat. Young animals that are bred for meat purposes in a short time, cannot form a concentration of protein because protein formation is a long-term biological process. Faster formation of protein with feed input is not possible. Butchers who to cut out water-containing muscles with little fat, market this as lean meat, and it is consumed accordingly.

However, the term "lean" suggests gaunt or thin. How can such an animal be capable of providing meat if it is fattened? It is extremely difficult to breed animals with little fat. If production cannot fulfill the increasing demand, then it will be manipulated and profit-oriented science provides sufficient possibilities, regardless of the consequences, to achieve this goal. Commercially available pharmaceutical products such as clenbuterol, ractopamine and salbutamol are used as feed additives to reduce the fat content in animals, thus providing plenty of lean meat. All of these drugs are extremely harmful; carcinogenic and can cause fatalities. (EU bans Brazilian pork supply, agrarheute.com, 26.03.2010/German.China.Org.Cn 15th and 03, 2011, lean meat products)

From a slaughtered animal, the muscles with less fat are cut out as lean meat. But what happens to the rest of the animal body? Who will eat the fat, the fatty meat and the remaining carcass? It is turned into industrial raw materials such as industrial grease, minced meat and animal food. Lean meat production produces more waste-meat than ordinary meat with fat. A better source of lean meat would be if one let the animals live to the end of their natural life span. In old age there is a natural weight loss and the remaining meat is considered to be better than the water, fat and pollutants in lean meat from a young animal.

The minced meat phenomenon

Minced meat production is a laborious process for modern industry. All body parts of an animal are separated from the bones and then systematically minced until the cell membranes are destroyed. In this condition, the meat becomes very tender and it is prone to bacterial contamination. Carbon monoxide and Carbon dioxide are used in packaged minced meat to sustain red colour and to inhibit contamination. Minced meat can be prepared in countless different ways. Minced meat is a raw material, which similar to molten iron, can take on any form. Industrially-produced minced meat can be adapted to country-specific cuisine or globally-known recipes. Modern minced meat is so flexible that it can be a raw material for almost any meat product. There is no limit to what modern cooking methods can do with minced meat and there is no other product which can compete with it.

New kinds of minced meat can be produced from different types of meat. A mixture of meat from omnivores such as pigs with the meat from herbivores such as cattle results in minced meat with a new flavor. New varieties of minced meat can be produced from these different commercially-available types of meat. The mixing proportions such as 1:3 or 3:1 may also result in new products. In many ways it is more profitable to mince the whole animal and sell this as raw material. Fresh minced meat cannot be kept long, but finished products such as ground beef patties, Koetbuller, cevapcici, kofte or hamburgers, can be stored in airtight packaging or refrigerated for a longer period of time.

In future, meat consumption will be much more oriented on finished products made of minced meat; which is boneless and cheaply-produced meat. In addition to this, a greater variety of meat is on offer, with different tastes, and there is easier access to meat as a staple food. However, this leads to problems such as nutritional diseases and contagious diseases in livestock.

Meat as food for children

The cubs of feline predators such as tigers are exclusively fed on milk until the second month of life, because their milk teeth start to grow slowly from the second week after birth. In the case of humans, the baby teeth start to grow after 6 months, but the baby food industry provides pureed beef for infants well before the development of primary teeth.

Small children do not generally like meat - a natural phenomenon, because their teeth are not yet fully developed. But they love processed meat products such as sausages, ham, salami or mince patties. These meat products contain similar flavor enhancers, predominantly in the form of salt and fat. Without such ingredients no child would choose to eat such meat products. Meat products manipulated with salt form a major part of food produced for children. This development is particularly apparent in industrialized countries. Despite very advanced pediatrics, the number of modern childhood diseases in these countries is extremely high. Different types of allergies, obesity, and many infectious diseases are common health problems in children. Experts look for the causes of new children's diseases in the polluted environment, poor nutrition, lack of mobility but very rarely in meat consumption. In this case the term "malnutrition" is a result of children consuming finished products such as sweets and sugary drinks.

A new debate on child nutrition must be undertaken, because future generations will have fewer possibilities to decide which food products are suitable for them and which are not. Mothers' milk is largely rejected, while infants and children are fed with meat and other processed foods. First of all, trade and industry benefits from the production of unsuitable baby food and secondly, and the pharmaceutical industry benefits likewise with their remedies for various childhood diseases. Ultimately the child becomes a permanent customer. For the rest of their lives they will suffer from certain diseases and spend their incomes buying treatments and remedies.

Meat in old age

Old age should be a pleasant time, when one is calm, serene, wise and essentially free of anger and greed. One has time to think about the past, to experience the present and accept the future. One can be happy about one's offspring, the changing environment and the good deeds that one has done in the past. Society respects aging people who complain little and have left an impressive and exemplary life behind. This time is considered to be retirement, time to rest. But in modern society this natural phenomenon has become something different.

When a person gets older, he gradually starts to develop health problems. These complaints in various forms are often accomapnied by pain or discomfort. People may become more vulnerable in their advanced years and eventually require nursing. This stage in their lives becomes the most troublesome time in their whole lives. People who have worked all their lives, earned money, paid taxes and given birth to children are regarded as a burden to taxpayers, or they are criticized for swallowing up most of the health care budget. All of these allegations are true. Everyone grows old provided one is lucky not to die young. But why must people suffer in old age as they do in developed countries? The older mammals in nature do not suffer from the so-called diseases of old age; they die suddenly of old age.

The four stages of life are childhood, adolescence, adulthood and old age. The diet in childhood and adolescence can be influenced by parents. The diet in adulthood is influenced in some respects by childhood and adolescence. What happens in adulthood determines nutrition and health in old age. The causes of most age-related diseases such as diabetes, heart attacks and cardiovascular diseases have their origin mainly in the increased consumption of meat. If older people would suddenly give up eating meat, they would not be freed from their ailments. But experience has shown that older people live more comfortably and suffer less if they do not consume food of animal origin. Staple food of plant origin, fresh fruit and vegetables, fresh water and exercise can lead to a satisfying life in old age. People who are over 50 years old have a better chance of spending a pleasant autumn of life if they stop eating meat. It is an individual decision, but a better future can be patterned only in the present. The homes for elderly people in industrialized countries provide their residents with a lot of meat and other animal products. Changing the diet in those nursing homes from meat to plant food is currently not possible, but would be exactly the right course to take.

Members of the Jain faith do not eat meat and suffer less from old age-related ailments. Donald Watson, the founder of Vegan Society was 95 years old when he died. He did not suffer from age-related diseases. He consumed no meat and needed no medication. In the last ten years of his life, Watson climbed the high mountains in north of England and he died quietly in the bosom of his family.

Hypothesis on body parasites

The food consumed also benefits numerous parasites that live inside the body. If a pet, like a cat, is fed with a certain pet food for a long period of time, it will usually reject any other pet food. The cat will survive for some days without food and then it will start to eat the new food. The following question arises: Is it the cat itself that refuses the new food or is it the body parasites that motivate the cat to do so? A mammal is host to innumerable large and small parasites. These parasites may control the host for their own interests. Many parasites affect the nervous system in a host, and the host acts according to those influences. As for example the small liver fluke (Dicrocoelium dendriticum) attains the abdominal cavity of an ant through food intake, control its nervous system and conducts it to the food source for many herbivores, where they can reach a final host. Greed or craving for meat can be caused by the presence of certain body parasites.

If a habitual meat eater suddenly gives up meat, he suffers greatly despite the availabilty of other foods. It can be assumed that the parasites present in this person's body are suffering from lack of meat and constantly craving it. A climax is reached, and if meat-eating starts again, then more than the usual amount of meat is consumed. However, if the temptation to start eating meat again is resisted, then the craving for meat gradually disappears and after a certain time the parasites in the body will lose their influence or get used to the host's new way of eating. Utterance like "I cannot live without meat" can be a state manipulated by parasites. This way the person can be a slave to his body parasites and eat what they want. Until now, there has been very little research into how parasites respond to very different diet types, or whether they can influence the cravings of their hosts. Perhaps drug abuse can also affect the nervous system by causing 'addicted' parasites to require more and more drugs.

Foreign influence from outside is an another chapter. In a football match the home team usually plays much better than the away team. The encouragement of the fans affects the nervous system of the home team and yet if the away team could steady thier nerves or exercise more self-control, they could win the game. Cultural influences, buying power and special offers encourage the consumption of meat. If at a banquet the majority of the guests do not eat meat or a celebrity guest does not eat meat, many other habitual meat eaters would follow their example. This type of external influence can lead to an appetite for or aversion to eating meat.

Possible return of cannibalism

As an omnivore, early man was also a cannibal. The world history of hunger and malnutrition is full of instances of cannibalism, many of which have been documented in times of war. However, cannibalism also occurs in times of peace even though other foods are available, as well as in times of famine. Jonathan Swift, the Irish writer and author of the famous novel

Gulliver's Travels (1726) wrote in another book - A Modest Proposal (1729) in which he proposes that the poor in Ireland should sell their one-year-old children for consumption by the rich, as this would provide an income for the parents. However, cannibalism is not just a fantasy. It has been practiced in various forms throughout history. Swift's idea of selling the flesh of children was later played down as a satire. But the idea of eating a one-year-old child finds a parallel in today's pork production. In both cases, the child and the pig are only allowed to reach two percent of their natural life span. In a period of time where food was very scarce, one could hardly allow jokes about it. Piero Camporesi has written an overview of sacred and profane cannibalism and the state of nutrition from the Renaissance to modern times. Hunger prevailed without interruption and people were not in a position to refuse unethical food (Camporesi, Bread of Dreams, pp. 40-55, Chicago 1996).

Passionate meat eaters usually seek meat with new flavors. Religious communities specify which meat from which animals may be eaten. However, breaking these cultural rules is not unknown. The most widely-banned meats are beef and pork, which are consumed by meat-eaters in large quantity. Interestingly the number of beef and pork consumers is not low in such religions where the consumption of those meats are forbidden. The variety of commercially-produced meat is very limited. Chicken, pork, beef and lamb make up about 95% of world meat production. It is boring to always eat the same kinds of meat. In some Far Eastern countries where religious influences are not very strong, a large variety of meats are on offer. The desire for meat consumption shows no bounds, and experimenting with new flavors has become a lifestyle. People try rats, monkeys, cats, snakes, elephants, songbirds, migratory birds, lizards, whales or dogs and their search for new meat flavors continues.

The afterbirth, also called the placenta, is relished as meat by many people. Numerous examples of this can be found on the internet. It is the only type of meat from a human which can be harvested without hurting the person. The placenta is a temporary organ in the uterus that develops simultaneously with the fetus and consists mainly of blood and mucus. The primary function of this organ is to absorb nutrients from the mother's body and then supply the fetus through the umbilical cord. This procedure ends with the birth of the child, after which the placenta is excreted. Now to the question: why is this organ called the placenta? In Latin, placenta means cake. Placentophagy, in which the placenta is eaten, is a widespread practice in nature. Food was scarce in ancient times and anything palatable was ingested. The term placenta, meaning cake, imparts legitimization to treat it as something edible.

Cannibalism is a social taboo and the eating the placenta breaks this taboo. If a person can eat a placenta, he is theoretically also able to consume the fetus which is connected to the placenta. Thus the consumption of a miscarriage, a premature birth or a stillbirth would be considered as legitimate. The claim that the placenta is a delicacy, that it is nutritious and has healing powers is used as justification for eating it. Nutrients and remedies are present everywhere in varying amounts. If the placenta was a delicacy, many commercially available placentas of sows and cows would be for sale on supermarket shelves. It is obvious that a new path leading to cannibalism is being taken through the human placenta.

"Ripe Eating", the ceremonial consumption of tribal elders, was a term used in some traditional societies. According to some traditions, the elders were selected, a festival was celebrated and finally a great feast made from these elders was prepared. The victims were convinced that the folk who consumed them would welcome them into their bodies. This so-called genetic predisposition may result from the idea that during his lifetime, a person can be a body donor, so that after his death, his body can be released to be eaten, similar to an organ donor, who releases his organs for medical purposes. A continuation of the use of human body parts such as the placenta or the consumption of other body parts for healing purposes may some day lead to cannibalism, which might be legalized, as in ancient times. The growing world population, lack of food and booming consumption of meat can be compared with the phenomenon of a locust infestation, in which in the final stage, the animals eat each other.

The Nauru syndrome

The Nauru syndrome was observed among the native people on the island of Nauru between 1970 to 2000, a period of 30 years. The Nauru syndrome describes a sudden change in the customary nourishment status caused by sudden wealth, which results in nutritional diseases. Nauru is a small and isolated island with a small population. Economic change was uniform and the entire population was directly affected by this change. For this reason, the Nauru phenomenon is regarded as unique and it can be used as an example of similar developments.

Lying close to the equator, the Pacific island of Nauru, with an area of only 21 square kilometers, has been settled for a long time. This very secluded coral island, covered in bird droppings, has barely any fertile land. No mammals besides humans have lived there and there are no sources of fresh water. Although there is a little rain, food crops such as coconuts and panda nuts thrive successfully. In the past, the people on Nauru did not eat any meat from mammals. Limited fish species from the coral waters, the occasional frigate bird and nuts made up their usual diet. When in 1798 English whalers came close to the shores of Nauru, they called it "Pleasant Island" because the islanders, who approached the whaling ship in small boats, demonstrated no hostility. This was certainly due to the fact that the people there hardly ate meat and so they were not aggressive by nature.

The bird droppings on Nauru, also called guano, contain a very high amount of phosphate. Phosphate, which can be sold as a valuable fertilizer for modern agriculture, was intensively mined in Nauru from the second half of 20th century and exported to developed countries. In 1968 Nauru became independent and in 1970 the phosphate mining was transfered to the local government. The population of Nauru was only 4000. Because of the lucrative trade in guano, the per capita income suddenly became the second highest in the world, after Saudi Arabia. Indeed the per capita income of Nauru was actually the highest in the world because Nauru had no monarchy nor any existing social hierarchy and so the whole population owned these revenues. The standard of living rose in a short time to one of the highest in the world. Guest workers were used to mine the harsh and harmful phosphate. Due to the abundance of wealth, the locals needed to do hardly any work and that small barren island became a kind of luxury hotel.

The enormous buying power enabled the Nauruans to have an abundant and varied diet. Value foods such as fish, meat, eggs and dairy products were imported and consumed excessively. In under two decades, the population of Nauru became the fattest in the world and over 90% of older people suffered from diabetes, heart attacks and other fatal food-related diseases. First-grade free medical care with the best medicines and the best treatment could not help to reduce this dangerous tendency. By the millennium, the phosphate mining was nearly exhausted and at the same time the short-lived wealth of the islanders ended. What remains is a destroyed landscape and the ruined health of older Nauruan.

Meat for the working class

Food is the primary means to force people to become dependent and the quality of the food is more important than having enough food. Meat is generally considered to be the best, tastiest, most valuable and most expensive food. Meat was also the best way to recruit soldiers. For example, just before the Second World War, the soldiers in Switzerland were provided with six times as much meat as the average Swiss civilian (Kollath, W; Textbook of Hygiene, Vol II, p 250, Stuttgart 1949). Until the beginning of the second half of 20th Century, the working class of developed nations could hardly afford any meat. The economically growing industrial nations saw the need and the importance of the working class - the backbone of the economy. Meat was offered to the working class, to keep them obedient so that they would not go on strike or become dissatisfied.

A fast-growing supply of meat was found in chickens and pigs. With plenty of food these two species can be locked up in a limited space. Chickens and pigs eat almost constantly in captivity and very quickly develop ample soft muscle tissue, which is called meat. These two kinds of meat have dominated the diet of the working class. Because beef and lamb are more expensive, they were only eaten occasionally. The mass production of cheap meat brought satisfaction and economic growth increased. Meat prices went down, the range increased and people ate more meat than ever before. Within a few decades, meat became the staple food of the working class. Thus, the working class remained faithful, loyal and grateful.

Within less than two decades, the Nauru-syndrome started appearing among the working class. Nutritional diseases or diseases of affluence became the focal point of health care. Medical expenditure rose rapidly, and absenteeism at work due to illness, early retirement and early death became more frequent. The pharmaceutical industry, which benefited from this situation, could not help the sufferers. Instead, they extended their life-span with painkillers. Pensioners, who had paid their pension contributions for three decades, were not able to enjoy their retirement because their health problems had robbed them of their joy of life.

The availability of cheap food of animal origin and its over-consumption is rampant throughout the world. The Nauru-syndrome is ubiquitous. Within two to three decades, the working class is at first well-nourished but then the population goes into decline. Few of these surviving workers become pensioners, but are commonly referred to as the inefficient and unproductive percentage of the population.

China's meat fatality

Despite advanced and intensive agriculture, throughout its entire history, China has been plagued by hunger and malnutrition. In the struggle to survive, the Chinese discovered more and more foods of plant and animal origin, which led to an endless diversity in their diet. The staple food of the Chinese people was rice, accompanied by a variety of plant foods and only rarely did their diet include fish or meat. In the 1960s the per capita meat consumption was only 6 kg per year; four decades later, it rose to more than 62 kg - an increase of over 900%. The current annual increase in meat consumption is more than 3%, and in a few decades China will achieve the per capita meat consumption of Euro-American standards.

Traditionally, Chinese meat consumption involved eating almost the whole animal, and each limb and body part was a delicacy. Everything edible from an animal was prepared with spices and salt. As long as meat was expensive, the custom of eating the whole animal was common almost everywhere in the world. As meat became cheaper, the consumer showed a preference for certain parts of the animal's body, which were considered a delicacy. Soon the Chinese will abandon their traditional style of meat consumption. Instead, they will prefer the soft breast, fillet and steak, and the rest will be processed as waste, animal feed or as fertilizer.

The phenomenon of the new type of meat consumption parallels the consumption of opium in the 18th and 19th

Centuries. The British imported opium to China as payment for the tea exported to England. Within a few decades, opium consumption became a major problem. The opium addicts were unable to work, became sick and died at an early age. Thereupon, the Chinese emperor forbade the importation of opium into China. As a result, the Opium War broke out (1839-1842), and in spite of being a great empire, China surrendered to the British. Instead of consuming imported opium, the Chinese then began to cultivate opium themselves. By the end of the Second World War, China remained the largest opium producer and consumer in the world. This dark chapter in Chinese history is full of misery and famine, until Mao Zedong from 1949 onwards managed to curb the production and consumption of opium.

The one-child policy and later women's employment brought the population growth almost to a point of stagnation and the accompanying economic growth claimed more and more workers. As incomes have increased so the production of meat has gone up and simultaneously meat has become widely available and cheaper. The Chinese have never before consumed as much meat and other high-quality animal food products as today. So far no nation has been spared from the consequences of this kind of excessive meat consumption. Modern nutritional diseases now affect populations of all ages. China has the highest number of diabetics in the world (The world's biggest diabetes epidemic, China org.cn, 29.08.2011/BBC World Service: The cost of obesity, 24.03.2014) and traditional Chinese medicine is not in a position to heal modern nutritional diseases. Traditional Chinese medicine had never had to deal with modern nutritional diseases before, because they did not exist.

The People´s Republic of China is in the midst of the Nauru-syndrome. All major pharmaceutical companies in the world are already present on Chinese soil and participate in the lucrative business of treating nutritional diseases. A drastic reduction in the consumption of meat and reverting to traditional foods is the only healthy option for China. The Chinese leadership must realize that the excessive consumption of meat is far more dangerous than opium.

The new rich of the world

Never before has there been so much money and wealth on earth as at the present time. The end of communism started spreading the market economy. The market economy and globalization in turn produced the nouveau riche. Wealth is used to buy food and the most important part of a meal contains products of animal origin. People who have consumed little of these products are slowly beginning to become familiar with them. The more wealth there is, the more dairy products, fish, eggs and meat are regarded as ordinary food. This is where the Nauru-syndrome sets in. Within a few years, people grow fatter and this is perceived by the outside world as a sign of prosperity. At a certain point, such people start to feel uncomfortable, but the thought of their purchasing power, a doctor's visit and better drugs soothes their fears. Their discomfort intensifies as time goes on. Visits to the doctor become more frequent and the range of drugs they are forced to take, becomes more important than daily food. However, good drugs, better doctors and expensive hospitals do not always help and a person's life is doomed.

Wealth means consumerism. Active consumption relates to food whereas passive consumption relates to anything that is not food and has little to do with health. New wealth enables changes to be made but it is unfortunate that these opportunities are often missed. Instead of developing healthy eating habits, people stuff themselves with the carcasses of dead animals and become sick. Money is a physical and emotional potential and it should not be used to damage ones own health.

The role of the pharmaceutical industry

The gradual industrialization throughout the 19th Century facilited the marketing of products which were manufactured in the pharmacies. From the time when vaccines, antiseptics, anaesthetics, antibiotics or insulin became available at the beginning of the 20th Century, the industrial production of medicines followed. The First World War, followed by the Second World War mobilized the worldwide pharmaceutical industry to produce drugs of every kind. Similar to the weapons industry, the drug industry became an inseparable part of the machinery of war.

When the two world wars were finally over and the surviving soldiers were healed, the war-oriented and state-funded pharmaceutical industry was threatened with bankruptcy. Many businesses did not survive the crisis, while others, who found a new orientation by applying their drugs to the civilian population, became successful. From the mid-20th Century the global pharmaceutical industry devoted itself entirely to conquering the civilian market and in a few decades became one of the largest industries in the world. They grew, made increasing profit and put more and more new drugs on the market.

The modern pharmaceutical industry had discovered a weakness in human beings. This weakness, called gluttony or feasting, has become the greatest boon for the drug industries. The more wealth there is, the more nutritional diseases there are. These diseases are also called diseases of affluence. The pharmaceutical industry is closely monitoring the spread of these diseases and producing appropriate remedies. The peak of meat consumption in the developed world was also the peak of drug consumption. The demand for medical services in industrialized countries reached its peak in the new millennium.

Thanks to globalization, the pharmaceutical industry has been able to spread to emerging markets wherever new purchasing power has appeared. Growing income has resulted in an increase in the consumption of foods of animal origin and private hospitals have become a new industry in emerging countries. Nutritional diseases are no longer typical of the traditional industrialized countries, but increasing purchasing power has spread this phenomenon to the entire world population.

Meat also allows passive earning for the pharmaceutical industry, which occurs in relation to zoonoses. The pharmaceutical industry has set aside a significant budget for research and development, and lauches remedies against animal epidemics and pandemics. Many of these drugs are preventive vaccines, injections or oral agents which are also taken as prophylactics. All these medicines originated only because of meat, which benefits solely the pharmaceutical industry.

The pharmaceutical industry no longer has the moral duty to heal or to help mankind. Instead profit has become more important than nature and humanity. Otherwise they would have explicitly warned against the consumption of meat and other food products of animal origin. This negligent act of the pharmaceutical industry is an act of self-deception and it can be considered as medical anarchism. A world full of diseases and sick people will remain the legacy of the pharmaceutical industry if they do not revert to their duties in time.

Health insurance may give a helping hand

The health insurance in the organized countries has a long tradition. This type of insurance expands gradually in developing and emerging countries. People with regular income conclude such insurances and benefit from the medical ministrations. The largest expenditure of health insurances are hospitalization, doctor fees and cost of medication. Because these expenditures constantly increase the insurance companies also raise premiums which leads to unpleasant disputes between insurers and the insured.

If health insurances would elaborately inform the insured that food of animal origin is harmful, the number of patients would reduce gradually. Subsequently, the insurance premiums could be reduced. The insurance companies should develop an appropriate form of questionaries and checkups, so that diet of the insured can be monitored on a voluntary basis. The vegan people should then pay the lowest contribution.

The interests of health insurance companies are contrary to the interests of the pharmaceutical industry, doctors and hospitals. The more sick people, the more loss to the insurance companies, but more income for the pharmaceutical industry. There is hope for a better future if the health insurance companies will take the matter of nutrition seriously and restructure their entire insurance concept.

Meat and the food processing industry

The food processing industry plays a very important role in food supply. With rising prosperity, more and more finished and processed foods are consumed. The food industry makes it possible to eat many foodstuffs that are usually not eaten through processing. Filleted fish, kitchen-ready steaks or minced meat without bones are the most popular industrial products. Fish and meat are made available by the food industry in all forms from raw to ready to eat. This is one of the major reasons for the increase in the consumption of meat.

The food industry has two main objectives: First of all, profit and expansion and secondly, more and more healthy and affluent consumers. A quick profit has consequences. To ensure healthy and wealthy consumers, the food industry must supply long-term healthy food. Otherwise the consumer becomes ill, and he then stays away from supermarkets. To attract new customers, more and more attractive, tasty and cheap foods are offered. But the number of consumers will not continue to increase because the new customers too become ill. This is a vicious circle set in motion by the food industry.

"The most valuable asset of a nation is its population" is a well-known saying from the 17th Century. This means not only the size but also the quality of the population. The health of the population is more significant than its wealth. A healthy nation is valuable not only to the state but also for trade, industry and services.

The new nutritional diseases spread quickly where the modern food industry has expanded successfully. The main foodstuffs that the industry delivers are of animal origin with basic substances such as fish, meat, milk and eggs. All of these high quality products are not ordinary human foods which can be consumed several times a day without any consequences to one's health.

Attractive, tasty, healthy and cheap food can be produced from plant products in an infinite variety. However, there is a lack of imagination, desire, ambition and interest in producers and consumers. The short-lived greed of the food industry and the consumer not only destroys their own future, but also damages the environment to an as yet unknown extent. In a world where hardly anyone is willing to work in agriculture or in any form of food production, people have no other choice than to hand over the whole tasks to large-scale industries. The industry pays attention to hygiene, storage, price, attractiveness, taste, flavour and the use of the approved methods of production. Food products are handled and delivered every few seconds. Modern man consumes finished foodstuffs more than ever before in human history. Even if he knew the full details of the modality of production, he cannot reject the industrially prepared food because he has no alternative. This dilemma has no end and the blame lies not only with the food industry, but with the consumers themselves and the existing system.

Restaurants

Restaurants in traditional industrial countries are suffering from the declining number of customers. Many restaurants have already closed and for many others it is just a question of time. Those restaurants that are still doing well are located in tourist areas. People coming into a foreign town, whether as a tourist or a traveler, have no opportunity to prepare their own meals. So, they are forced to visit restaurants and not necessarily because they want to sample the local cuisine. To keep within their budget, they often eat cheap food. Most travel illnesses are directly linked to restaurant visits and the consumption of foreign food. The reason why the traditional catering industry is in danger, is not the declining population or reduced purchasing power, but rather the poor health status of restaurant visitors, caused by the unhealthy food served there.

When meat was scarce and expensive, many people went to restaurants. The evidence for this assertion is the existence of numerous traditional pubs, restaurants and taverns. Every small town and every neighborhood had restaurants. They were well-patronized, otherwise this trade could not have devloped. Also the exclusive facilities of many of these restaurants are a reminder of their glorious past. The amount of fish and meat served was very small, and it was tough and bony, so it was not possible to satisfy one's appetite. The main food which satisfied restaurant visitors, were of plant origin.

Since the beginning of the last quarter of the 20th Century, meat has become cheaper and cheaper. Restaurants have changed their usual menus to a meat-menus and the traditional staple food of plant origin have been demoted or become a negligible accompaniment. The food offered was given new

51

names which were derived from a particular fish or meat. Restaurant owners provided their visitors with as much meat as possible and in the 1980s the restaurant attendance reached its historic peak. Themed meat-eating evenings were launched, which attracted restaurant visitors in droves, in small cliques or larger groups, to enjoy an orgy of meat. After a short time, the number of visitors decreased steadily. Again and again, new restaurants have opened, with new designs and new menus but many of them have had to close.

To make a profit, the surviving restaurants had to come up with new ideas. The new source of profit has become not the food but the drinks. The restaurants attract customers by offering cheap meat, which is prepared with an excessive amount of salt so that they can make their profit from the sale of expensive drinks. Most of these drinks are alcoholic or sugary, which is an unhealthy combination with meat as a staple food. So, they attract customers with meat but they make their profit with alcohol. In order to increase the number of visitors to restaurants, other ideas have been tried out. Special events are organized, which are not typical for a restaurant, for example, large monitors for football broadcasts.

Another way in which restaurants seek to acquire customers, is by only allowing their toilets to be used by paying guests. Such business ideas, invented to increase the profits of restaurants, are not a promising future for the catering industry, which has allowed itself to be diverted from its original objective of serving nutritious meals.

Most restaurants sell only meat, and the choice for non-meat eaters is limited. Sometimes they offer vegetarian guests a salad, soup or a pancake as an alternative to meat and treat them as diabetics or eccentric. The food offered by modern innkeepers is primarily limited to meat, and the skill of cooking food derived from plants remains alien to them.

The limited knowledge of modern restaurateurs recalls the events of the great famine caused by the potato crop failure in the years 1845-1848 in Ireland. After the introduction of the potato in Ireland, it became more and more popular. Because potatoes are easy to cultivate and more productive, other foods were neglected. When suddenly the potato crop failed, a great famine resulted. The knowledge about the cultivation and care of other food crops, which had been widely grown and eaten before potatoes were imported, had been lost. One million Irish people died during this famine, two million emigrated and the population of Ireland shrank from eight million to five million. For many of the two million Irish emigrants who embarked in ramshackle vessels on a voyage to the new continent of America, the Atlantic Ocean became their graves.

The culinary knowledge of innkeepers is identical to that of housewives, where ordinary meals are based only on meat. Whether it is breakfast, lunch, dinner, a snack or a feast, nothing is possible without meat. In the modern diet, the aroma and flavor of meat as well as the Umami-feeling are essential. The restaurant owners claim that they are influenced by what the customer demands and accordingly they serve mainly meat. But they in fact are the people who use meat as a cheap trick to make a quick profit.

Animal diseases, animal epdemics, contaminated meat, rotten meat and various meat scandals often bring the consumption of meat to a halt. Restaurants suffer from lack of visitors, consumers lose their appetitie for meat and wait after such incidents for the reappearance of safe meat. Like a drug, meat remains an inseparable companion of modern life, and a compromise with plant food seems to be hard to imagine.

An authentic and reliable business idea would be a purely plant-based restaurant. Such restaurants would offer a variety of basic foods such as potatoes, rice, maize and wheat, and side dishes of legumes made from the limitless selection of vegetables and fruits. Along with this type of food, tap water certified as clean would be available at no cost. A culinary cuisine from around the world would banish the memory of meals based on meat and the food would be cheaper than conventional food made from animal carcasses. Food hygiene and health concerns would no longer be a problem.

The status of man and passive food poisoning

During a case of active food poisoning, the effects take place immediately or within a short period of time. The cause may be an intentional assault or a microbial contamination. Passive poisoning is a long-term process, but the causes are the same as for active food poisoning. Long-term health complaints develop from passive food poisoning which can be classified as different diseases, including nutritional diseases.

Although many people live in a modern society, many social functions are still defined as either men or woman's work. Producing food and trading in food is a man's job, and preparing food at home is the task of a woman. Different professions are also divided according to sex. Men go to work to earn a living and women prepare daily meals. An exception is very rare; even working women often have to perform the same tasks as traditional housewives on top of their salaried jobs.

Cooking as a profession has little to do with preparing food at home on a daily basis. A professional cook does a job to earn a livelihood and this cannot be compared to a housewife who cooks at home. This male profession was created by faith-based cultures. Because women were not allowed to work in public, this task was left to men. For traditional celebrations, a cook is usually hired, not because a woman would not be physically able to prepare food for hundreds of participants in the celebration, but because cooking on such a large scale cannot be done in a closed kitchen, but must be done outside in presence of many people.

Statements such as "the best cooks are men" were invented by men. A meal prepared by the best cook in the world has a psychological effect on the eater rather than the quality of the meal. The famous French cook Paul Bocuse was chosen as the "Chef of the Century". The 85-year-old Paul Bocuse once said: "nouvelle cuisine is: nothing on the plate, all on the bill" (Culinary Institue of America, New York, 01.03.2011).

Most men cannot cook. The only type of cooking that men can manage is grilling meat outside on an open fire. Many men do not think much of the art of cooking and cannot imagine doing it themselves, but they all want to eat a wide variety of good, tasty food. The best housewife is judged according to her cooking skills. She greatly appreciates praise for her culinary efforts, but she does not want to be accused by her husband of always cooking the same.

Restaurant owners use the same methods as housewives to ensure their food always tastes good. It is not only skillful preparation that makes a tasty meal; the ingredients are important too. Using ingredients which originate from animals is the simplest way for housewives to achieve recognition. Housewives give their men a lot of meat and other animal products, for which they earn praise and recognition. Ironically, this traditional reassurance by men leads to slow food poisoning. Men are the ones who suffer most from nutritional diseases, get heart attacks or require replacement organs, all because of those housewives who cook so well for them. Women do not teach their sons to cook, do not involve their husbands in food preparation and they claim "Men have no place in the kitchen."

This division of labor functioned in traditional societies, in which there were hardly any nutritional diseases, because the daily diet consisted mainly of vegetables. This has all changed because of the abundance of meat. Most men in a modern household are supplied in the same way as the animals reared for meat in a barn. These animals are slaughtered at an early age and therefore the symptoms of eating disorders have not had time to develop. But if they lived longer and continued to be fed, they would develop the same diseases as humans.

Modern civilization should no longer have gender-based roles. Women could change the world if they gave up their monopoly of the kitchen and cooking and did not pander to the tastes of men. The interior of the body is very sensitive and vulnerable. The nourishment that we introduce into our bodies must always be monitored and the best way to control food intake is if one is involved in its preparation. Many famous men are amateur cooks. They have learned this skill not necessarily to avoid possible poisoning, but they know how to manage their own health with regard to nutrition. Taste and satiation should not always be related to purchasing power.

Suffering caused by eating meat

Dental Complaints

Human teeth are not suited to chewing meat. The tedious chewing process burdens the teeth, wears them down quickly so that they become weak, loose and fall out prematurely. Even the teeth of carnivores such as members of the cat family become damaged due to eating meat. Compared to herbivores like the horse, a tiger loses its teeth much more quickly. A very old horse retains its teeth, but a very old tiger has only a few of its teeth left. Before the implementation of modern dentistry, many people in adulthood ate only the soft innards of an animal such as the liver, kidney or heart. Because of dental problems many people eat only minced meat which can be swallowed without chewing. Meat soup was originally intended only for those who, because of bad teeth, could not eat meat in any other form.

Chewing meat damages baby teeth at an early age and then later the permanent teeth are also affected. The frequent chewing of meat puts stress on the root of a tooth and long-term exposure weakens teeth, so that they have to be pulled out or fall out by themselves. Regular brushing and regular visits to the dentist are precautionary measures to protect teeth. Toothpaste is the most widely-used body care product of the modern age. Meat consumption and dental care have undergone a parallel development. So, the more meat is consumed, the greater the need for dental treatment. Industrialized nations have the highest per capita meat consumption in the world and at the same time these countries have the highest per capita expenditure on dental care.

Teeth are sensitive body parts and toothache is one of the most painful ailments in life. Toothache caused by caries could not be treated until the first half of the 20th Century. The only treatment that was available for toothache was extraction. This task was often performed by a barber, blacksmith or healer. A loose and brittle tooth can be pulled out easily but not a permanent tooth. There was no dental anesthesia available when pulling a permanent tooth.

The development of anesthesia and the dental anaesthesia used in human medicine is a chapter in itself. The backbone of modern dentistry is the invention of dental anaesthesia. Lidocaine, the first anesthetic used in dental treatment was developed in 1943 and was marketed from 1948 onwards. Although the pioneer product Procain or Novocain was invented in the year 1905, it took decades to produce a variety of dental anesthetics in dentistry (Rahn, Rainer / Ball, Benedict, Local Anesthesia in Dentistry pp. 6, Frankfurt 2001). Until the successful use of dental anaesthesia, mankind suffered from a phobia of going to the dentist. This, perhaps genetically-driven fear of dental pain and anxiety about dental treatment, is still present in many people today.

In the second half of the 20th Century, modern dentistry achieved its breakthrough. Rotten and tarnished teeth could be treated, pulled out, pierced or replaced painlessly. After the invention of dental anaesthesia, prosthetic dentistry had been improved so greatly that it sometimes allowed minor procedures to be carried out without anesthesia. But the focus has always been on dental anaesthesia and it is used for the smallest complaints. This ground-breaking development in dentistry has allowed for a carefree attitude to the consumption of meat in the developed world. Meat tastes best if it is chewed and not just swallowed down. Weak teeth are treated continuously by being drilled, ground, filled, replaced or implanted. Furthermore, bridges are built, crowns are fixed and and dentures are fitted in order to allow people to chew meat. In fact, the human mouth is not adapted to eating meat , but modern dentistry makes it possible, thereby preparing the patients for treatment with conventional medicine.

Poor teeth are often thought to be a result of an improper diet and lack of dental maintainance. Sugary food products are regarded as unhealthy and the cause of tooth decay. But meat, which can consume and weaken the teeth more quickly, is being ignored. The highest aim of dentistry is the preservation of a person's natural teeth. The field of dentistry should be honest about the fact that chewing meat is injurous to teeth. If this warning is heeded in childhood, despite having caries caused by sugar, the teeth will not need to be extracted or replaced. The only disadvantage will be the financial decline of the entire dental industry.

Eye diseases

According to a study carried out at Oxford University, people who consume meat are more vulnerable to eye diseases such as cataracts than people with a meatless diet. A total of 27,670 volunteers, who were over 40 years old and did not suffer from diabetes, participated in this study. The largest group at risk of cataract formation were the excessive meat eaters. In second place were the normal meat-eaters. This was followed by people who ate less meat, fish or were vegetarians. Bringing up the rear were the vegans. In fact, the risk of vegans developing cataracts was 40% lower than in the excessive meat eaters (Diet, vegetarianism, and cataract risk. American Journal of Clinical Nutrition, March 2011). Meat consumption is thus partly linked to eye problems and eye-related medical expenditure.

Meat and dementia

Obliviousness or forgetfulness was usually an accompanying symptom of many aging people and they were provided with simple household tasks. But during recent decades it gradually becoming a pestilence with much perplexity. Until in the 70s of the 20th Century disease like dementia was broadly ignored. Evidence of this claim are those different kind of literature of that period of time. The term dementia in those days was explained as feebleness of mind or schizophrenia whereas other modern diseases were described extensively.

Modern scientists have found that the accumulation of amyloid fibrils in various organs like brain, liver, heart, kidney and pancreas cause disorders like Alzheimer, Parkinson and different kind of fatal diseases. Some 30 proteins form amyloid fibrils in filament structure which accumulate in those organs. They are insoluble and resistant to enzymes which can dissolve unwanted proteins. Amyloid fibrils are found as plaques in the brain of dementia patients. Aging people specially above the age of 60 years are endangered to such kind of diseases. The origin of those toxic proteins is unknown and until now there is no remedy to get rid of or to wash up those evil and sticking proteins.

Dementia is a global disease but people of rich and industrial countries with higher income are more vulnerable to it.

According to the report of World Health Organization 58% of dementia patients are in under-developing and developing countries and only 42% are in rich and developed countries with higher income. But those rich and developed countries make only 14% of the world population. This means that 42% of dementia is allotted to this minority of global inhabitants. The reason could be the excessive and regular consumption of animal products such as meat. Because the prime recommendations of dementia prevention is to eat fruits, vegetables and many other products of plant origin and avoid products which are rich in cholesterol and saturated fatty acids. Interestingly those cholesterol and saturated fatty acids are the true ingredients of meat and other animal products. Other advises of dementia prevention such as spiritual, physical and social occupations can be termed as less helpful. Because plenty of people always occupied with such tasks also suffer from dementia. Moreover there is no scientific evidence that the plaques of amyloid fibrils in the brain can be reduced with such spiritual or physical activities. Reducing as well as abandoning the consumption of animal products such as meat as early as possible can be a better way to prevent or avoid dementia.

Zoonosis

Zoonoses are diseases that are transmitted from animals to humans. Contagious diseases are transmitted by viruses, bacteria, parasites, spores, fungi, and prions. More than 850 diseases that are transmitted to humans are considered to be zoonoses. The transmission of zoonotic diseases occurs mainly through pets and farm animals that come into contact with humans and rarely through wild species. The most common and worst epidemics or pandemic-like spread of zoonoses are bird-flu, swine-flu and other flu-like diseases. The viruses of these deadly diseases proliferate in the animal body and are then transferred to healthy people. The worldwide zoonoses during the last three decades have caused more economic and physical damage than any other natural disaster in this period. 75% of all new diseases that have been transmitted to human in the past 10 years are from the pathogens of livestock and food from animal origin (WHO, Zoonoses and veterinary public health, Geneva, June 2012).

Meat, virus and infectious diseases

Because viruses do not possess any visible metabolism, they are assumed as non-living substance. But their body formation consists of organic ingredients like DNA, RNA, lipids and proteins and they reproduce or multiply. Metabolism is a process in animals that consume food in coarse material form, yield energy after digestion and discharge unwanted materials as excrement. The plant world does not possess any digestive system and the food intake is based on two different ways- photosynthesis and via roots. Within the process of photosynthesis solar energy is converted to food energy, but the roots beneath the earth search for nutrients in molecular level such as nitrogen, phosphorus or potash. On the contrary a virus can reproduce only inside a host cell. Perhaps the viruses have an undefined metabolic scheme for energy absorption by abiding in a host cell where they multiply and eventually catapult out of the cell.

There are billions of different viruses in the nature which have their own host in the world of plants, animals as well as in microorganism. Only some hazardous viruses for human and animal are discussed within this scrutiny. Some contagious diseases of recent decades are bird flu, swine flu or SARS epidemic. An interesting observation on those virus diseases is that they were active in those high level meat consuming regions of the world. These viruses usually attacked people that were accustomed to meat consumption.

Also low level meat consuming regions may suffer from such viruses if the meat is from some particular origin. The existence of the Ebola virus has been detected in several species of fruit bats (Megachiroptera). The fruit bats are directly accused for spreading this virus in humans and primates. Although fruit bats are mammals, they do not develop symptoms of the disease caused by the Ebola virus. Fruit bats feed on fruits, nectar and flowers and no information of foods of animal origin are present in their somatic cells. So they are not affected by the Ebola virus as it does not multiply. Instead the virus itself uses fruit bats for transportation most likely to reach a host. But in case of bird flu bred poultry is vulnerable to such diseases, because they are fed with different fodder of animal origin. It is conceivable that such evil viruses like bird flu, swine flu, SARS or Ebola can only propagate in hosts who consume meat and other foods of animal origin.

General health problems

A diet without meat is more beneficial to health than a diet with meat. Good health secures a good income by reducing medical expenses. If parts of a dead foreign body pass through a living body, the biomass of the foreign body leaves tracks in the living body. Molecular compounds as well as larger parasites of the foreign body may find a new home in the new body. This assertion of a biological fingerprint left by foreign bodies is neither a superstition nor an unscientific theory. Proponents of meat consumption see only the beneficial ingredients of meat. They firmly believe that heat treatment destroys the harmful content of meat, whether organic or inorganic, leaving only the beneficial content. This idea is not only naive but it is also unscientific. It has been proven scientifically that a number of toxins, proteins and molecular compounds such as ciguatoxins, aflatoxins or prions are heat-resistant. Breeding chemicals, antibiotics or growth hormones are also largely resistant to heat and are transferred to the foreign body. It may be that many other heat-resistant molecular compounds in an animal body have not yet been discovered and they may play a role in the transfer of biological information.

Animal fat, which after its consumption does not leave the body of the consumer, has great potential to affect health negatively. Also, the cholesterol content in meat is very high and causes numerous diseases. In the last three decades, countless scientific studies have proven that meat consumption increases the risk of many life-threatening diseases. Diseases are physical problems that cause pain and disability. The naming of a disease is often more significant than the complaint itself. There are no good diseases and there is nothing good about any kind of physical pain a person has to suffer.

Insomnia and meat as a staple food

A sleeping disorder is not a disease but may be classified as a disease if it becomes a constant companion. It is estimated that over 20% of people suffer from sleep disorders. As with all complaints, the causes are always different. Sleep disorders may be caused by noise, light, pain and anxiety but they may also have their origin in what is consumed. Stimulants such as tea, coffee, tobacco, drugs, alcohol or medication can cause insomnia. If all these things are not the cause of a sleep disorder, it may be caused by diet. The study of sleep is an excellent scientific field. Although sleep therapies and sleeping pills are flooding the pharmaceutical market, 25% of the population in industrialized countries, where meat consumption is very high, suffer from sleep disorders.

Eating, drinking (water) and sleeping are the main requirements of living beings. When food intake is completed, sleep follows for the restoration of the body. The type of food eaten, which assist sleep, has an important impact on it. If the body is not adapted to certain foods, the wrong foods may lead to insomnia. As there is a lack of anatomical support, the human digestive tract has more problems with foods of animal origin and therefore may sleep less at night than carnivores. The staple food of a lion is meat, and a lion sleeps 18 hours a day. But the sleep of a lion is very light and the slightest disturbance, even a leaf falling from a tree, can wake it. In addition, all predators are nocturnal animals. Hardly any big carnivorous mammals sleep throughout the night.

Humans are not animals, so why is animal testing carried out to benefit mankind? The study of the behavior of animals in their natural environment is more scientific than animal experiments in a laboratory. An experiment on a shackled monkey in a laboratory will certainly produce a different result from experiments on monkeys in the wild. Herbivores sleep longer, eat more and take almost no naps. In contrast, naps are very common among carnivorous mammals. Abundant plant food and fresh drinking water give a better quality of sleep than carcasses in the abdomen.

Organ donation and meat consumption

So far, no studies are available as to whether excessive meat consumers need organ replacements more often than consumers who eat little or no meat. However, countries with a very high per capita meat consumption are the pioneers in organ requirement, organ donation and organ transplantation. The most frequently damaged organs are the kidneys, the heart, the liver and the lungs. Only the kidneys and the liver can be donated by a living donor. Governments and many other institutions make public relations to attract donors. Such a possible donor carries a donor-card and is afraid of a fatal accident, perhaps someone is waiting to benefit from his organs. What a creepy and macabre life imagination.

According to evidence, the human liver cannot detoxify vitamin A (retinol), which the liver of carnivorous species can. The function of the human kidneys is different from that of carnivorous animals, as the urine of carnivores is very concentrated whereas the urine of humans is only moderately concentrated (whale.to/a/comp.Mills, MR, The Comparative Anatomy of Eating, 07.06.2012). Meat increases the risk of coronary heart disease, which can lead to heart failure. Diseases caused by eating meat may also damage sensitive organs that may need to be replaced. Of course human organs are not always available in a replacement market. People travel to India to buy a new taxi for a poor but healthy taxi driver and in return get one of his two kidneys.

During the last three decades a global trade in organs has developed. The most affected people are the poorest in the poorer countries, who can hardly afford meat or other foods of animal origin. Officially the trade in organs is regarded as illegal, but the practice continues in the shadow economy. The importation of a prohibited merchandise such as weapons or drugs can be monitored and the goods seized, but not a transplanted organ. Police have discovered million-dollar scandals, where over a long period of time, hundreds of kindneys were extracted from poor workers in India and sold for 1250 U.S. Dollars each. These kidneys were then implanted into customers from rich nations (Reuters; Kidney racket scandal in Gurgaon shocks India, New Delhi, Mon Jan 28, 2008 3:26 IST). Why is India the centre of the market in organs? Could it be because the per capita meat consumption in India is less than 5 kg and therefore Indians have healthier organs?

The construction of modern hospitals in the developing world is a new industrial boom. The rich benefit from these new hospitals. Many of these hospitals advertise for patients from around the world. Retired people from industrialized countries are lured to these hospitals with promises of better care. The opportunities here for finding a potential organ doner are greater than by being on the waiting list at home. Missing children and people who simply vanish are a worldwide phenomenon, but this problem is more common in developing countries than in industrialized countries. Purchasing power is not the only reason for buying other people's organs. A serious debate on morality and ethics is missing, to decide whether organ transplantation is a scientific miracle or a medical monstrosity.

Instead of just looking for organ donors, the governments of rich industrialized nations should also take health measures in the field of nutrition. A successful and well-enforced measure is the ban on smoking in all public, enclosed spaces. Excessive sugar intake and alcohol consumption are also harmful to a person's health. Tobacco, sugar and alcohol can damage a person's health, but meat damages not only health, but world peace and nature as well, to an unknown degree. To ensure the reduction of excessive meat consumption, the government does not need to get involved in expensive and strenuous amounts of work. As the first step, they should cancel stock-breeding subsidies completely and in addition, introduce a tax on meat, like the tax on tobacco.

Meat consumption and the human body

Hands: Human hands are adapted to picking, collecting, digging, touching, searching, feeling, climbing, swimming, painting, pointing, signing or shaking hands. No other animal species possesses such skilful hands as humans. But without aid, these ingenious hands are not adapted to catching terrestrial or aquatic animals in their natural habitats. Hands are not designed for killing a mammal, tearing the meat or chopping it up into bite-size pieces. Suitable hands for catching animals, are paws and claws, and not human hands.

Mouth, teeth and the digestive tract: Predators, also called carnivores, possess canines and incisors. The frontal canines are used primarily for hunting purposes and the scissor-like incisors further back help to cut the meat into pieces. The specially-adapted skull of predators allows the mouth to open very wide to catch the prey and eventually to dismember it. The meat is simply swallowed, without chewing. Humans do not possess the anatomical and physiological characteristics of carnivorous animals. The front teeth and back molars of humans are not suited to chewing meat, either raw or cooked. Meat cannot be swallowed down by humans without chewing, as meat-eating animals do. The human oesophagus would not allow it and as a result the person would suffocate. The oesophagus of an adult human is 25-30 cm long, but has a diameter of only 2 cm. The long narrow tube does not allow a human to swallow even a thin slice of ham without chewing. Meat is a product that causes difficulties for the mouth, teeth, throat and oesophagus of humans. If, despite all these obstacles, meat is consumed, other complications may be triggered in the body.

The digestive tract: The human digestive tract is completely different from that of a carnivorous predator. Digestive enzymes, the size of the small intestine and the colon of predators are not identical to the human digestive tract. Carnivorous animals have developed the ability to digest meat and not plant cellulose. The human digestive tract is identical to that of herbivores such as cattle, monkeys and deer. Man will not succeed in a Darwinian attempt to change his digestive tract in order to digest meat like a carnivore, because this would also affect his external appearance.

Protection and attack mechanisms of the body: All vertebrates possess anatomical defenses. These mechanism can also be used to attack. Larger herbivores use their horns, hooves and tusks as weapons. Carnivores use their paws, claws and canines. Man is a large vertebrate and in relation to his body size and in comparison to other species, he is anatomically very poorly armed. Human fingers, toes and nails are weak, and can be damaged from even the smallest injury. The mouth has no special adaptation which allows it to attack or defend himself. The human mouth opening is too small for attack purposes and is only big enough to allow communicate and the ingestion of food. According to the theories of evolution, the human body could have evolved a body full of scales or long horns on his skull. But even the remains of prehistoric man show no physical defenses. The belief in the Abrahamic faith that God created Adam and Eve, and then expelled them from heaven to earth as a punishment, is in this context easier to accept, rather than Darwin's evolutionary theory.

Primitive man started using tools to defend himself and to help him collect food. The object in his hand, be it a stick or a stone, was more efficient than his own fingernails and teeth, and so unused biomechanisms receded. These tools became transformed in the course of time into true weapons. The modern firearm is a development of the original objects thrown. The natural adaptations in anmials are used by herbivores for defensive purposes and by hunting predators for offensive purposes. Man is an omnivor and used his weapons for two purposes - defensive and offensive. Since the beginning of the modern era, man has been practicing the double use of his weapons.

The use of weapons has enabled man to play a greater role in nature. The psyche of the opponent was affected by the sight of a weapon. Man, who had a stick in his hand, could scare a physically superior animal, and feel safe. Man's knowledge of defense and offense allowed him to develop dominion over animals. Even dangerous animals like the lion or tiger were afraid of a simple stick because the injuries in carnivores are far more devastating than in other physically large herbivores. The develpoment of defense using foreign objects caused the biological defense mechanisms to fade away. This claim can be proven because every person is capable of increasing a certain body part through years of training, be it a lip or ear lobe. Man can train his hand to break stones or to take glowing charcoal into his mouth. Weapons allowed humans to have access to meat consumption, freed the body from unpleasant pain, gave him security and at the same time his body became more vulnerable. This also led to a separation from nature.

Meat and gender

Male: Meat is a gender-oriented food which is preferred by males. From antiquity to the present day, it is mostly men who are involved in hunting, breeding, trading, and butchering. Men can easily cut up the meat of a slaughtered animal and then roast larger chunks on an outdoor grill. Even in modern society this kind of primitive male hobby is still widespread. Hunting, butchering and meat-eating was a symbol of manhood. With this act of violence, men could seize leadership. Meat was a product of a struggle and a successful struggle allowed the acquisition of meat. In this sense meat is not simply a kind of food, but also a symbol of power. With meat, men dominated and ruled. The dominant species among animals are the predators and the dominant gender is often the male. Man observed this behavior in carnivores and tried to imitate them.

Female: Women are not skillful hunters, nor are they successful cattle traders. They breed a smaller number of domestic animals, rear them affectionately and seldom slaughter any of them. This characteristic is not because they are too weak or frightened, but because they do not have the urge to kill an animal. It is an interesting observation that no other food is consumed by men so greedily and not by women. Usually it is women who are responsible for cooking and the meat prepared by women is mostly of a smaller quantity. This means that they prepare meat along with other food. Interestingly women provide men with more meat than they consume themselves.

Women suffer more from meat allergies and excessive meat consumption causes them more health problems than men. Furthermore, women suffer far more from insomnia in industrialized countries than men (DGSM; The sleep of women, p 3, Berlin, 2011). This is because meat in industrial countries is consumed as a staple food and the female body cannot tolerate meat as well as the male body. It is mostly women who abstain from eating meat. They suffer less from dental problems, high blood pressure or heart attacks. Why do women seem to have a natural antipathy towards meat? The simple answer is the message rooted in a woman's instinct that she is not carrying a predator.

Chapter II: Meat production

The theory of critical grain

Grain or cereal is the staple food of the majority of the world's population. Its cultivation is hard, tedious and the meagre crop is highly valued. However, grain is handled carelessly if the supply exceeds the demand. When there is a surplus of grain production, it is usually sold. If the price of grain goes down, the grain is used to produce milk, eggs and meat by feeding it to farm animals. If this conversion results in more profit than selling the grain for direct consumption, the grain is then used as animal feed.

In subsistence economy, people consume rice, corn or wheat as a staple food several times a day. About 600 to 800 grams of grain is needed to provide 60 to 80 percent of a person's daily food requirements. The remaining 20 to 40 percent of the daily diet comes from food supplements of various plant and animal origin. The per capita share of staple foods accounts for 220 kg to 290 kg of grain a year. The upper limit of grain consumption is 300 kg and the lowest is 200 kg. Below the 200 kg limit, the supply is insufficient, and this can lead to malnutrition. If the per capita consumption of cereal exceeds the 300 kg mark, there will be a surplus of grain. However, 300 kg of cereal per capita cannot be achieved with traditional farming methods. The premises of traditional agriculture are scarce arable land and the use of traditional farming methods. Under these conditions, the per capita grain production will usually remain below the 300 kg limit.

The per capita cereal production resulting from traditional and modern farming methods is between under 200 kg and over 1500 kg. All countries with higher grain production increasingly produce and consume foods of animal origin. Countries like the

USA, Canada or Australia produce more than 1000 kg of grain per capita of the population, while the per capita meat consumption in these countries is more than 100 kg per year. Countries with less than 300 kg per capita cereal consumption, such as India and Bangladesh, produce 5 and 4 kg of meat per capita per year respectively. The increase in meat production in relation to grain production can be seen in the case of Indonesia. Indonesia produces just over 350 kg of cereal per capita and as a result the per capita consumption of meat has gone up to 13 kg per year (FAOSTAT).

The total world production of cereals is around 2.5 billion tonnes. This is a very good average of 350 kg of cereals per capita for a world population of 7 billion. However, almost 50% of this grain is used as animal feed. Hence, the per capita cereal average drops below the 160 kg limit, thus affecting a large part of the world's population, excluding the wealthy consumers. A general ban on the use of grain as animal feed will not occur, because industry-oriented crop production is primarily feed-oriented. A ban on the use of grain as feed would cause a decline in cereal production and the lack of food would remain. Other ways of eradicating hunger should be sought, rather than the production of abundant cereals.

No association between meat consumption and world hunger

Meat consumption causes hunger. Many scientists, development organizations, environmentalists and economists claim that the excessive consumption of meat is responsible for hunger. These institutions and individuals all assert that modern meat production uses large areas of fertile soil and fresh water resources for fodder production. This fodder is often cereals or legumes, which could be consumed by humans as food. For example half of the world's grain harvest is fed to animals to produce meat. If humans would reduce their excessive consumption of meat or give up eating meat entirely, there would be no more hunger. The land could be used to grow food or feed grains, which could be consumed as food. The increase in the world's population is also cited as a key cause of global hunger and malnutrition. These kinds of abstract ideas about hunger have arisen due to the absence of surveys on the history of hunger, demographic transitions and a neglect of the market economy.

When the population of the world was less than one billion, there was very little grain, there was hardly any meat and the world suffered from hunger and malnutrition, a fact which cannot be denied. When one million people died of starvation in 1848 in Ireland, there was neither grain nor meat, and when 40 million people died of starvation in 1962 in the People's Republic of China, the Chinese population was only 400 million. Five decades later, the population of China has grown to 1.4 billion, but hardly anyone is hungry.

Fodder production is a market-based development. If the demand for meat declined, fodder production would also decrease accordingly. If the demand for food grain rose, the production of food grains would also increase. But what would be the decisive criteria which would help to fight hunger? It is only purchasing power that can combat or eradicate hunger and not just putting an end to meat consumption.

The eradication of hunger does not depend on a reduction in the consumption of meat, but rather on individual criteria such as elementary education, the education of women and female employment. No country in the world suffers from hunger and malnutrition where these three expectations are fulfilled. Rising meat consumption has a negative effect primarily on people's health and the environment, with disastrous consequences.

The most commonly-eaten types of meat

The animal world is very varied but man has restricted himself to only a few species, which provide him regularly with meat. This has been the case since the beginning of animal husbandry. Pigs, chickens, cattle, goats and sheep have been domesticated for a very long time and they supply over 94% of the meat consumed all over the world today. The future of these four species can be defined as biological meat machines and no other species will ever reach a similar level of importance.

Pig: Pigs are the largest source of meat in the world. Other than supplying man with meat, this species is of almost no other use to humans. Pigs give birth up to three times a year, producing altogether over 20 piglets and with adequate food supplies, these piglets reach a body weight of up to 150 kg in less than 6 months. Intensive pig farming requires very little space and as omnivores, these animals can convert all possible types of feed into meat. The demand for meat worldwide is tremendous and pigs have a limitless ability to meet this challenge.

1961	25
1992	72
2012	112

World pork production between the years 1961 and 2012 in millions of tonnes

In the space of 50 years, pork production increased from 25 to 112 million tonnes. There has been a huge growth in pig breeding, but the demand for pork is very high and the target to be met is still a long way away. Culturally, pork is not accepted everywhere and for religious reasons at least half of the world's population does not consume pork. Nevertheless, the global demand for meat is growing and the increase in per capita income will in turn benefit meat production. For mass consumption, meat cannot be produced in the laboratory (as is often claimed), and no other species is able to compete with the pig. The cultural rejection of pork is only apparent in active use, i.e if the consumer can assure himself first hand of the origin of the meat. Considered as a commodity, pork as a raw material has the biggest potential for the manufacture of meat products. Because consumers are neither breeders nor butchers, it is very difficult for them to analyze the meat content in the package. Pork offers the versatility needed to meet the urgent demands of the convenience food industry.

Chicken: Chickens have two uses: they lay eggs and produce meat. In intensive chicken farming, chickens are bred very successfully in very little space. Chickens are omnivores. Their constant feed efficiency enables rapid growth and meat formation of more than one kilogram in less than a month. In addition, when there is a decline in egg production, the layers are processed into meat.

1961	8
1992	39
2012	93

The world chicken meat production between 1961 and 2012 in millions of tonnes

There are almost no cultural taboos against eating meat from chickens. All you need to produce chicken is chicken feed and this will be in the foreground of food debates in the coming years.

Intensive chicken farming, which was only practiced in industrialized countries, gradually spread almost all over the world because intensive chicken farming is more economical than traditional methods of production. Until the 1960s, chicken was an expensive type of meat and people could rarely afford it. Nowadays, chicken is one of the cheapest meats, affordable even by low income groups.

The demand for chicken meat is worldwide. International trade barriers have fallen and the production facilities in North America, South America and Europe have a vast potential.

The disadvantages of chicken meat is the serious spread of H5N1, also known as bird flu. If H5N1 can be controlled successfully, global chicken production would experience an exponential growth.

Cattle: In intensive cattle breeding, the animals are used in three ways: they produce milk and meat, and supply leather. Extensive cattle farming is based on traditional mass herding. In mass herding, cattle are kept on large areas. They are primarily used for meat and leather. Traditional cattle farming is one of the oldest forms of animal husbandry. As draft animals, the animals were of great importance. Later, milk and meat production became important, but this depended on the cultural setting. In traditional cattle farming manure and leather are considered less important by-products. The water buffalo belongs to the same bovine family and therefore it is also counted as cattle.

Modern beef production is directly related to milk production. In both cases, leather is regarded as an important industrial raw material. Milk, meat and leather make cattle farming a lucrative business. Apart from Hindus and Jains, people of almost all cultures consume beef. In contrast, leather and cow's milk are commonly used and consumed by all cultures. The demand for cow's milk is increasing worldwide and cowhide has established itself as an important industrial raw material. At the same time, beef offers a great alternative to the fast-growing production of pig and chicken meat. The global cattle population will continue to grow in the coming decades. Compared to pigs and chickens, the number of cattle is not increasing as rapidly. This is due to traditional cattle farming in the so-called Third World countries and the increase in intensive cattle farming for milk production.

1961	28
1992	53
2012	64

World beef production between 1961 and 2012 in million of tonnes

Fodder and pasture are the basic requirements of the cattle industry. The dairy industry has no alternative but to raise cattle. In the year 2012 a total of 626 million tonnes of cow's milk were produced, a per capita average of 89 kg of milk per year for the entire world population. In the industrialized countries about 100 kg of milk per capita is consumed directly and more than 200 kg of milk per capita are consumed in the form of dairy products such as butter, cheese and yogurt. The trend in the demand for cattle is inclining towards milk products, which in the coming decades will be a tough challenge on the environment.

Theoretically, a world population of 7 billion needs, with 300 kg milk per capita in a year, a total of 2.1 billion tonnes of cow's milk. The cattle population in 2012, not counting water buffalo, was 1479 million animals, which delivered a total of 626 million tonnes of milk. On top of that, 97 million tonnes of water buffalo milk was produced. In order to achieve the goal of more than 2 billion tonnes of milk, the number of cattle has to be increased by more than three times. This means, a cattle population of over 5 billion animals is needed. In addition to this, the availability of feed must also increase more than threefold. In this calculation, all end products such as beef, leather and slurry must be included.

Goats and sheep: Sheep and goats belong to the group of goat-like animals that are the fourth-largest providers of meat and other products. These animals eat almost any plants, but they are difficult to confine and the meat and milk production in these species is very low. However, because of the abundance of food for these animals in nature, export-oriented sheep farming operates in countries such as Australia and New Zealand. Wool, goatskin and milk are among the most important by-products.

1961	6
1992	10
2012	14

World goat meat and lamb production between 1961 and 2012 in million of tonnes

Goats and sheep play a minor role in the supply of milk, meat and leather and their droppings cause few environmental problems. However, they eat the natural vegetation voraciously, which can cause major environmental disasters. Coincidentally, Australia´s dingo population has also increased. These wild dogs kill large numbers of sheep every year, and dingos are now the target of many Australian hunters.

Abundance of meat from wild game

There were two key developments caused by the decimation and extinction of predators of all kinds. First, the abundance of game meat and secondly the steady growth in the number of farm animals. Parallel to the killing of predators, other animals were hunted for food, as well as for trophies. From the mid 19th Century, the supply of game meat began to increase. Not only did hunters have access to game meat but the urban population could also acquire wild meat regularly at the butcher's. There were two major reasons for the abundance of wild meat. The first was the decimation and destruction of predators, so their prey such as deer, stag, wild boar etc. were able to proliferate freely. The second reason was the progressive efficiency of firearms, which resulted in new professional hunters who provided regular game meat for consumers. Since the end of the 19th Century up to the early 20th Century the hunting of wild animals in some areas was so extensive, that instead of eating the meat, people ate only the soft organs such as the brain, heart and liver. They threw away or abandoned the remaining carcass. Delicacies such as goose lever paté, oxtail soup etc. appeared during the period of abundance of game meat. Large flocks of wild geese were killed in their resting places and often instead of eating the tough meat only the soft liver was preferred. Another example: hunters killed large numbers of wild buffaloes, only to cut off the tail to make oxtail-soup and the rest of the carcasses were simply left to nature.

Game meat today

Despite the steady decline of wild animals, the traditional consumption of game meat has remained unchanged and in some regions it has even increased. The consumption of game meat is dependent on the food culture, in which different wild animals are consumed. According to the IUCN, many species are threatened with extinction only because of consumption by humans and many species have already become extinct. A worldwide study has revealed that 3 to 4 million tonnes of venison was consumed yearly during the first decade of this century (Trends in Ecology and Evolution, Voll. 18, No. 7, London 2003). Accoring to different food cultures, almost all kinds of wild animals are hunted and eaten.

Little songsbirds are eaten as delicacy – the better the songbird can sing more tasty its meat is. In the entire mediterrenian region and in some European countries migratory songbirds are caught with different cruel medthods to prepare a culinary meal and in sub-Saharan Africa bush-meat or ape-meat is consumed as ordinary food.

Smart-phones are smartest among other methods to inform customers about the catch or hunt of an wild animal. Poachers just take a picture, send it to their agents and those agents inform consumers including culinary restaurants. If a restaurant want to buy that kill, it means that the restaurant already informed several customers who would like to dine with it. It could be a hoary monster of the deep Southern Ocean, or steaks of a very heavy and strange moose from the virgin forests of Alaska, all of them taste different and those well-funded buyers think that it is their time and life to enjoy.

The steady population growth depletes forests and the habitats of wild animals. Lack of protection encourages the on-going consumption of game meat and economic interest ensures that this practice continues. Restaurants use larger amounts of game meat and there is a global trade in wild meat of almost all existing species. The global supply of wild meat is widely used in e-commerce. Even protected species are hunted and offered for sale. Most of these products are delivered to the consumer by home delivery. Wine producers market their wines as 'Delicious with game', new recipes showcase game meat dishes and even supermarkets have venison in their range. Although a lot of the game sold in supermarkets comes from the breeding of wild species, they foster the tendency to consume wild meat.

As an alternative to venison, meat from farm animals is recommended but this cannot be regarded as a proper solution. The breeding of farm animals may be threatened by predators and the extermination of predators would increases the number of wild herbivores that would eat up the plants. The decimation of wild herbivores by hunting would also bring no solution, because the balance between flora and fauna would be disturbed.

Meat and fodder from a global perspective

Animal husbandry means that animals do not have to forage for food; instead, the breeder alone is responsible for supplying the feed. Feeding and the domestication of wild animals are directly connected. If domesticated animals had continued to forage independently, the animals would have either strayed or been eaten by predators. Nomads wandered from place to place with their families and animals in search of fodder. In contrast, sedentary farmers could always supervise their animals. In both cases he was solely responsible for the supply of fodder. In other words, animal husbandry can be equated with fodder.

Everything that humans consume can also be regarded as fodder, and everything that humans do not consume because they do not like the taste or it disagrees with them, can also be considered as fodder. The worldwide production of over four billion tonnes of fodder per year are considered as very scarce. In this predicament people turned to wild flora, fauna and many other possible sources of fodder. Plants which have no acute toxic effects can be eaten directly. Leaves, branches, bark, fruit, seeds and the roots of wild plants belong to the non-toxic fodder source. These wild plants are harvested in nature without providing for the next generation. After exhausting these sources, a new source is selected. The AFRIS of FAO (Animal Feed Resources Information System) have documented over 700 potential fodder sources including feed for farmed fish. Most of these sources have been researched scientifically and have been released for feeding. Toxic plants can be rendered harmless with heat treatment, such as repeated boiling and pouring off the water each time until it is safe to use as fodder. If unpalatable, non-toxic plants are mixed with other feed or cooked, the smell and taste is altered and it can then be used as fodder. It should be noted that, with the exception of extremely poisonous plants, the rest of the plant world is available as fodder.

Cereals, the most important food for mankind, are increasingly fed to farm animals. In this type of resource conversion, up to 90% of the plant food is lost. For this reason alone, the average per capita grain consumption of rich nations is over five times higher than the average grain consumption of the world population as a whole. In addition to cereals, legumes such as peas, chickpeas, lentils and oil crops like canola and soy are the most important food sources increasingly used as fodder.

In Europe, for example, even peas, which were an important food for humans, are now increasingly fed to pigs.

The abundance of synthetic detergents and lubricants pushes the prices of vegetable oils down. Consequently, cheap fats and oils are used to feed farm animals. At the same, time large numbers of toxin-containing by-products and waste products derived from vegetable oils are also fed to farm animals. Fats and oils are rich in energy, they cause faster growth, they encourage gluttony and prevent the formation of dust in feed. Moreover, the addition of 7% fat to the diet of poultry and pigs reduces the feces in terms of volume by half (AFRIS).

Obtaining fodder from the animal world is fairly new. Dead animals from breeding, the waste produced from salughtering, inexpensive meat, inedible meat, bones, feathers, blood, egg shells, whey, etc. are converted to fodder. The breeding animals supply food for human consumption and at the same time their body waste is used as fodder and fed to other breeding animals. This is called "recycling". Over 7 million tonnes of fishmeal are added annually to pig, chicken and fish feed. To produce one kilogram of fishmeal more than five kilograms of fresh fish are required from the wild. The oceans are being overfished for these purposes.

In addition, a variety of animal species in nature are considered to be a potential source of animal feed. Rats are found around the globe. Dead rats can be readily converted into protein flour. However, the FAO has warned that the rat poison that is used for trapping purposes, must not be harmful to farm animals. Rat flour used in raising chickens has shown good results. According to fodder experts, snails are present worldwide in great numbers. Some giant snails on the African continent can reach a length of up to 30 cm. Snails are rich in protein and easy to harvest. Snails are boiled for 10 to 15 minutes and mixed with up to 20% of other feeds. The use of snail feed could reduce the use of fishmeal in poultry feed by up to 50%. Snail feed has also shown good results in tilapia farming. The FAO has suggested that the potential of snails as a global source of fodder is enormous. Numerous frogs and toads live in many humid areas of the world. Those animals are also a valuable source of protein. Frogs are easy to harvest after dusk. Frogmeal and feed made from toads can be used for raising chickens and pigs, and in fish farming.

Huge numbers of squids live in certain waters. Squids are a high-quality source of protein which could easily replace fishmeal. Squidmeal is already being used very successfully in chicken and shrimp breeding. The oyster industry catches large numbers of starfish and transforms them into protein flour. This starfish flour contains toxic thiaminase and therefore the addition of 5% for chicken and pig feed should not be exceeded. Numerous other small animals that are present in some regions can be processed as animal feed. Insects such as grasshoppers, cicadas, beetles etc. are rich in protein that can also be used as animal feed. Earthworms, caterpillars, larvae, etc could also be used in a similar way.

Under the term "miscellaneous" many other substances can be classified as animal feed. Some examples of these are waste paper, chemicals, minerals, corals, leather or the excrement of factory farmed animals. In the broadest sense, all products of an organic and inorganic origin could be transformed into feed. Paper is of plant origin and is considered to be a fodder commodity. However, color used in printing contains heavy metals such as lead, but so far no significant negative effects on animals which were fed with a compound feed of waste paper have been observed. Waste paper is mainly used to feed ruminants. Wood contains plant cellulose that can be digested by ruminants. Wood waste can be softened by milling and heat treatment such as boiling, and then mixed with other feed and fed to ruminants.

Animal leather consists mostly of collagen, a glue-like protein substance. The waste strips of the leather industry are used as animal feed. Leather processing uses toxins such as chromium (Cr), and therefore the chromium content should not exceed 2.75% in feed made from leather waste. Milled leather has been used successfully in chicken feed as a substitute for fishmeal and in pig feed as a substitute for soy. The feces of pigs and chickens include 25% and 30% crude proteins. These are treated with heat, which removes pathogens and nematode eggs, and then dried and mixed with other feed or for the production of feed pellets. The feces of ruminants also include crude proteins, but these are processed as cattle feed. So, cattle have to get used to eating their own feces in their daily fodder. In contrast, there is no problem for chickens and pigs if their own feces are present in their feed. A 2 kg chicken produces 800 grams of feces per week, a large cow weighing 650 kg produces 150 kg and a pig of 45 kg produces 22 kg. The excrement of breeding animals are considered to be valuable feed resources. This use of excrement is referred to as an alternative disposal of animal feces. In fact, it is a convenient source of animal feed, in which the waste proteins are converted into edible proteins.

The larger the animal husbandry, the greater the demand for feed. In their search for new sources of feed, human beings have pushed the natural as well as the ethical boundaries further back. This results in the total exploitation of plants and animals. The use of animal excrement and animal carcasses as feed is also an offense against ethical values. The only thing that was not mentioned as a feed source is the use of human remains. The animals are forced to consume their own flesh and own feces, because man regards milk, eggs, fish and meat to be essential food. In a global market it is extremely difficult to trace the source of food of animal origin. Even a living animal shows neither its origin nor the manner of feeding. A farmer cannot guarantee that his animals have received clean feed if he is producing fish, meat, milk or eggs for the global market.

Goat breeding and desertification

A desert is an area with no or very few plants. Cactus, acacia, grass and shrub plants, which can survive with very little water, grow in oases or on the edge of the desert. The plants of the desert are barely edible for human, but animals such as goats eat them and provide a livelihood for desert inhabitants.

The population growth followed by economic growth required increasing income from livestock breeding. Extended animal husbandry in arid areas requires larger feed areas in the desert and on the edge of the desert. As a result, the borders of the desert are increasingly eaten bare, causing the desert to expand relentlessly. This phenomenon is particularly apparent in the African and Asian desert world where in the last three decades desertification was allowed to proceed undisturbed while meat production continued to grow. The Sahelian zone is one of the most sensitive desert regions in the world. Precipitation is extremely low, fertile farmland is rare, the population is growing and animal breeding, especially goat farming, is regarded as being the most important and viable enterprise. The following table shows the conditions of rapid growth of goat breeding in the Sahelian zone.

Countries	1961	1991	2013
Burkina Faso	2	7	13
Chad	2	3	7
Mali	4	7	19
Mauritania	3	4	6
Niger	5	7	14
Nigeria	1	26	58
Senegal	1	3	5

Goats in millions of heads of selected countries in the Sahelian zone

Source: FAOSTAT, Rome 2015

During this period the population more than doubled. The rapid growth in the number of farm animals was not due to increasing meat consumption. On the contrary, animal breeding served predominantly as a major source of income. The statistics of consumption of goat of selected breeding countries are dubious. For instance, the largest export earner of Mauritania between 2007 and 2009 was the export of meat meal (FAOSTATT). Carcass meal and meat meal are two different products. Carcass meal from different species is produced from slaughterhouse waste and dead animals, and this is mainly found in animal feed or as fertilizer for food cultivation. In contrast, meat meal is a product of one particular farm animal such as cattle, sheep, pigs or goats. The modern use of meat meal is a topic in itself. Meat meal is used as fish bait, as pet food and it is the raw material for the manufacture of different meat products. The meat of slaughtered animals is cooked, dried, ground and packed in bags. It is durable, lightweight and convenient for storage and transportation. Sausages, meatballs or minced meat sauces can be made from any kind of meat meal.

The poor animal breeders of Sahelian zone can hardly afford to eat meat themselves. The animals are usually sold at the age of one year and vegetable staples like maize, wheat, millet or rice are purchased from the proceeds. What happens to those animals remains almost hidden. On the other hand, the internet trade in meat meal is booming. The breeders of the Sahelian zone grow no feed, nor do they buy fodder through international trade. As an alternative to crop production the UN recommended livestock breeding in these arid areas, without thinking about the problem of feeding the animals. Subsequently, the UN convened a desert conference in order to discuss why the deserts are expanding. Abstract phenomena such as climate change are then be held responsible for global warming and desertification, while animal breeding and meat production are ignored.

Meat in the Arctic Circle

It is not known why people settled in extremely cold regions such as north of the Arctic Circle. Perhaps they fled north to escape attackers, perhaps they were cut off during an ice age or perhaps it was due to the abundance of food. It is also not known how long people have been living in these icy regions. Investigation is impossible due to a lack of evidence such as stones, wood or metal. But these people lived and hunted using traditional methods. Modernization has introduced firearms of every kind, motorized snow vehicles, motor boats, helicopters, warm clothes, electricity and satellite connections, so that today they can more easily hunt animals in the ice. Every person has the right to live where he wants, but not at any price. It can be assumed that in the past, the number of polar residents and the number of species living there were both very small. It is absurd to hunt the limited species using the most modern methods and destroy them in this sensitive part of the world where no plants grow.

The inhabitants of the Arctic Circle do not build igloos anymore, because they no longer follow the traditional way of life. They live in industrially-made dwellings, have purchasing power and hunt in the fragile wilderness. It makes no sense for people who live in these icy wastes to be permanently fed on imported food. The inhabitants of the polar regions should either give up modern hunting methods or migrate voluntarily to the south. The governments of those countries which lie within the Arctic Circle would be responsible for carrying out the re-settlement and giving them financial support for their new beginning.

Reindeer husbandry

Reindeer husbandry in the Arctic Circle was a reliable basis for the traditional economy on which the subsistence of many of the northern dwellers was built. Unfortunately this tradition has been introduced into the world market. Reindeer meat in any form - fresh, frozen, canned or processed is offered worldwide on the internet and also traded conventionally. It is even available in numerous brands of dog food. Statistics on reindeer meat are not yet available, but trading in reindeer meat is so profitable that even helicopters are deployed to round up the reindeer herds in the wild.

Reindeer breeding cannot be compared with any other livestock. Because of the poor or non-existent infrastructure, it is virtually impossible to supply fodder to the scattered herds of reindeer. Moreover, the animals are wild to semi-wild and cannot be domesticated or confined. The ever-increasing reindeer herds eat the sensitive lichen, moss-like vegetation, as well as grass, branches and leaves and so contribute to the desertification of the polar regions. The environmental damage caused by modern reindeer husbandry has not yet come to public attention, but once the damage becomes visible, it will certainly be very difficult or too late to rehabilitate these regions. By then the reindeer industry will be well-established and the impact on the environment will be regarded as normal. The businesses, jobs and economic growth will then be more important than the environement. This phenomenon of reindeer herding is identical to the goat farming on the edge of deserts, which is responsible for the expansion of the desert areas. If one day there is no more vegetation above the Arctic Circle, the profit-oriented scientists will declare this line to be the northernmost treeline and backdate the previous wilderness to an event before the Ice Age.

In coming decades, meat from the polar region will probably be advertised as the cleanest natural meat, and affluent consumers, perhaps even desert dwellers with purchasing power, will become regular consumers. A variety of products made of reindeer leather, like jackets, boots, gloves, etc. available today prove that reindeer meat has become an ordinary commodity. Powdered reindeer antlers are exported to Asia as an aphrodisiac. All in all, the potential of the reindeer herding is very promising because:

- the area inside the Arctic Circle is huge
- reindeer meat is considered to be very tender, lean and tasty
- reindeers grow almost without supervision
- reindeer forage in the wild
- reindeers do not produce slurry

The only thing the breeders have to do is to keep the natural predators of reindeer under control. They will surely hunt wolf, polar bear, brown bear and wolverine with helicopters from the air while herding the reindeer together.

Moose

Moose have been affected in much the same way as reindeer. Many northern countries see a great potential in moose meat, moose leather and moose hunting by tourists. Although the moose is a wild animal and will remain wild, its meat is traded internationally like an ordinary market product. The IUCN declared the moose not to be at risk because wolves, the natural enemies of the moose, were being systematically exterminated. One day, the polar region will be declared a wolf and bear-free area and the zoos of industrial cities will be proud to own some of those animals, which will then be in danger of extinction.

Milk in nature

Up to the time of birth, a baby or a calf is fed by the placenta through the umbilical cord. After birth, the function of the placenta ends and the breast or udder takes over the supply of nutrients. Milk is the liquid food for newborn mammals. Newborns, who cannot consume solid food, are supplied directly from the mother. In this case the food intake and food delivery takes place from body to body, without touching air, water or soil. This is the only natural way milk is supplied and the only role of milk in nature. For this reason the milk of other mammals is not regarded as a natural food for humans or other species.

Products which are meant for reproductive purposes may be a food source in nature. More than 99% of eggs, seeds and fruits are currently being used as food or feed. Even the blood of a living animal is a food source for several species. But there is no animal on earth, apart from man, which feeds on the milk of another animal. In South Asia there is a legend among the peasantry, which explains the decline in the milk yield of a dairy cow. It tells of the existence of a particular species of snake, about three feet long and striped yellow and black, which feeds exclusively on milk from a cow. This serpent, called Guwala-Saap or Milksnake (Lampropeltis triangulum), ties the rear legs of a dairy cow together and then sucks the milk from the udders. There is no proof of milk-drinking of this snake, nor has any veterinarian ever found any trace of such a snake bite on the udder of a cow. If made available, milk from a cow would be consumed by almost every living animal. Even a cow-eating tiger would take cow's milk from a bowl, but this does not mean that the animal world, with the exception of newborn mammals, consume milk. So, milk from a mammal should never be treated as food, because this phenomenon does not exist in nature.

Cow's milk has uniformly high levels of carbohydrates, fats and proteins and it has a water content of more than 87 percent. This in turn allows cow's milk to be used in many food products. All food products contain various valuable food ingredients such as vitamins, minerals, proteins and carbohydrates. Milk also contains many of these nutrients and it therefore regarded as a healthy food source. However, for every argument in favour of drinking milk as part of a balanced diet, there is a counter-argument based on evidence that dairy products can be harmful. The main argument used to support the consumption of milk is that the calcium content of milk is necessary for healthy bone structure but the counter-argument states that the calcium content in milk may cause osteoporosis and prostate cancer (Zentrum-der-Gesundheit, CH-8008 Zürich 2013). The calcium content in milk promotes the growth of bones in babies but not in adults.

The modern diet includes milk and dairy products as a staple food. Milk, butter, yogurt, cheese, curd, cream, etc. are consumed throughout the day from early morning until late at night. In addition to this, milk products are used as ingredients in other finished and semi-finished foods. Cheese is produced through an unnatural manipulation. First, a part of a calf's stomach is cut out, dried, ground and added to milk as rennet.

This macabre method of food manufacturing can be regarded as unethical. About 80% of the world population consume no cheese and refraining from eating cheese made with rennet, would hardly cause an upset in the human diet. It cannot be true that milk, a growth medium for newborns, can at the same time be a healthy food for all age groups.

Dairy cows are not allowed to live longer than 4 or 5 years because with age, the daily quantity of milk decreases. Slaughter and then the sale of the meat is the only available option. The feed input is calculated exactly along with the milk output. When the critical level is reached, at which the feed becomes more expensive than the income gained through the production of milk, the animal is slaughtered in order to achieve a quick profit. In this case, beef is a direct by-product of the dairy industry.

Milk production is directly related to calf production. The most desirable calves are the females because they in turn give birth to calves and at the same time produce milk. Male calves are not desired. Veterinary medicine cannot manipulate conception so that only female calves are born. According to the rule of nature, about half of the animals born are female and the other half male. The unproductive male calves are either killed immediately after birth or they are allowed to live for several weeks, so they can be sold as veal. Veal is advertised as containing lower levels of fat and more protein and veal is at its best between 4 and 6 weeks of a calf's life. In fact, the animals do not reach even 0.4% of their natural age and the meat consists of soft and sticky muscle with a water content of about 80%. This veal is a direct by-product of the dairy industry.

Over 90% of the milk used for food production purposes, is derived from cows. The more than 626 million tonnes of cow's milk is a by-product of fodder input and excrement output. The excrement, called slurry, contaminates the fertile soil and fresh water resources. Milk is slurry, and the slurry is fodder.

The egg as a natural food source

Eggs are an important food source in nature. The eggs of some species are consumed more than 99.9% as feed. Man as an omnivore eats almost all kinds of eggs. The anatomy of humans is well-suited to collecting large eggs. Fragile birds eggs can easily be stolen using the soft finger muscles and the eggs can be consumed fresh. The eggshells can easily be broken open. Birds eggs and turtle eggs were among the most important natural products sought by man, who climbed up tall trees and steep cliffs to plunder the nests, or he searched for turtle eggs in the sand.

Egg-collecting became a profession, like hunting or farming, and it required special skills. This included knowledge about the breeding seasons and the nests of various birds and reptiles. The egg-collector often exercised a kind of symbiosis, in which he drove off potential rivals for the eggs but he did not rob all the eggs from a single nest. The egg-collectors exchanged the eggs for food, for other goods and kept some of them for their own consumption. In the quiz question: "Which came first, the egg or the chicken", the answer is "the egg" because it is the egg which was collected by man. Only domesticated animals lay eggs in captivity and the domestication of bird species such as ducks and chickens took place only when humans became settled.

The egg was a product of symbiosis between humans and domesticated poultry. However, man has separated himself from this traditional dependence and put the chicken into small cages where it is given fodder. Egg production has increased rapidly and the growing demand for hen eggs show no limit. The following table provides a brief overview of egg-production between the years 1961 and 2012.

Year	Hen eggs in shell (number) in billions
1961	269
1992	766
2012	1250

Source: FAOSTAT, Rome 2015

For the 1250 billions of hen eggs in the year 2012, there were only 178 eggs per capita of the world population. In addition to their conventional use as a dietary supplement, eggs are used by the food industry and other industries as a raw material. In intensive poultry keeping the males are killed and disposed of immediately after hatching. But in the semi-intensive and extensive chicken farming the males may live for several weeks before being processed into meat. Laying hens are slaughtered according to the decline in egg production.

An egg has no functioning central nervous system and it is not sensitive to pain. Using this as an argument, eggs are consumed by some animal rights activists. Originally, an egg was a natural product and man's anatomical features allowed him to find and consume eggs. But industrial chicken farming means: the more hen eggs, the more chicken meat.

Meat and globalization

A series of trade agreements after the Second World War made trading easier. The GATT negotiations from the 1950s onward through the 1990s and the agreement between the European Union and Africa, Pacific Islands and the Caribbean Islands increasingly facilitated the global food trade. Despite all kinds of liberalization there was national protectionism, especially in the meat trade. But those days are gone. The trade agreement for a common global market was presented in 1995 in Marrakech in Morocco, and this finally lifted international trade barriers.

Under huge protests worldwide, the success of world trade was announced and all rushed to take advantage of it. Within a few years, world trade has turned into a common market place. The most interesting part is the trade in food products that are consumed every day and this trade has levelled out. Food products of plant origin are cheap, but require higher costs for transportation and storage. In contrast, food products of animal origin are not cheap, but the demand is very high, and the transportation and storage costs seem to be affordable. In particular, the global meat trade gained greatly in importance.

Industrial countries with ultra-modern agriculture like the USA, Canada, Australia, New Zealand and the EU began competing with each other in order to participate in the world meat trade. Also new emerging countries have started selling their meat products on the world market. Brazil, Uruguay, Argentina and the PR China are particularly advanced in the area of meat exports. The entire African continent, Asia, Eastern Europe and Latin America are the regions with imports for affluent consumers. The per capita income in these countries is gradually growing and the potential demand for meat is very large and growing all the time. The results of this world meat trade are very clear. Especially the fast-growing meat animals such as chickens and pigs are of great importance. The meat from these two animal species have captured the world meat market by a large margin compared to other meats. Pork and chicken form about 67% of the total world meat production. The following table provides an overview of world imports by major types of meat that came after the emergence of globalization:

Year	Chicken meat in tonnes	Pork in tonnes
1996	4.533.161	1.947.091
2000	5.929.314	2.839.785
2004	6.680.525	3.570.272
2008	9.778.471	4.511.523
2012	13.500.000	5.500.000

World importation of chicken and pork meat after globalization
(Source: FAO; Statistical Database, Rome 2015)

Within 16 years of globalization, the world trade in chicken meat grew by 198% and pork grew by over 182%. This is an annual average increase rate of 12,37% for chicken and 11.38% for pork. But the world population growth in this period was less than 1.2%. This enormous difference between the population growth and meat imports was due to the market liberalization and the rising incomes in emerging countries.

There is a great potential for expanding the meat trade in developing and newly-emerging countries. The exploitation of these markets is hindered not by the lack of purchasing power, but by the deficiency of refrigeration caused by lack of a reliable supply of electric power. There is a constant endeavour to expand the current network and find new sources of power. The construction of nuclear power plants and river dams for power generation are a new trend to achieve this goal. When this obstacle has been overcome, meat exportation from the United States will be more significant than the exportation of grain. Not only the USA, but also New Zealand, Australia, China, the European Union, Canada, Brazil, Uruguay and Argentina will participate competitively in this lucrative business and supply the world population with abundant meat.

The more modern the agriculture, the higher the production of chicken and pork. This has become possible due to favorable feed cultivation, government subsidies, the beneficial use of veterinary medicine, uninterrupted power supply and a developed transport infrastructure. It is cheaper to buy chicken or pork from the United States than to import feed from there. It is time-consuming to breed animals and then to process them into meat. Another advantage of simply importing the meat is that there is no threat of animal epidemics in the consumer country. Imported meat as a finished product only needs to be eaten. Industrialized countries can produce comparatively better quality meat, maintain punctuality and so lead the world in meat production and will continue to dominate the world meat market in the future.

For religious reasons, the production of pork may reach its limit sometime in the next few decades. In contrast, there is no visible limit in the demand for and production of chicken, because almost all cultures in the world consume chicken, which is inexpensive and can be produced at a high speed. Globalization connects meat consumption and the growth in trade and industry, which is disadvantageous for human health and ecology.

USA - The World Champion in meat production

The United States is the world champion in all areas of meat production. Nutritional diseases, caused by increased meat consumption have long been recognized as a public health hazard in the USA and the health issues are part of the unpleasant economic debate in the country. The temperate climate and the endless plains (Great Plains) in the centre, north and midwest of the USA has allowed the production of grain to an unlimited extent. A part of this area was originally called the Corn Belt or Grain Belt.

The increasing use of artificial fertilizers, the mechanization of agriculture and the widespread rail network has allowed the production and sale of grain to flourish. However, the amount of cereals produced, which once served as a staple food, exceeded the demand. As grain prices on the New York Stock Exchange on "Black Thursday" 24th October 1929, began to fall, the entire IPO collapsed within a short time. This led to the Great Depression in U.S. economic history. It took a long time to deal with the question of where to put or what to do with the U.S. grain. A global export of grain at that time was not feasible. Agricultural production declined and the economic crisis remained until the beginning of Second World War. After the war, as reconstruction began, people gradually discovered the beneficial use of grain as feed. From this point of time onwards, grain production started to increase continuously and the area of the Grain Belt in the midwest has become known under the new name of Hog Belt. The meat industry settled there where the corn grows and so the USA has gradually become the largest grain and meat producer in the world.

Meat produced in the USA is exported and consumed locally in excessive quantities. It is a profitable business for trade, industry and agriculture. However, it has had a devastating impact on the environment, which cannot be rectified with profits from meat production. Natural disasters in the form of tornadoes, droughts, bushfires, floods, heat waves or cold waves have become more frequent. Land erosion by the excessive use of fertilizer, salinization caused by irrigation or the pollution of drinking water as a result of intensive farming are also common side-effects. Moreover, the destruction of the vegetation, the extinction of wild animals, widespread zoonoses and resistant vermin are also the result of intensive agriculture. Finally, the health of the local population is also affected.

The USA produces 43 million tonnes of meat annually, over 93% of which is chicken, beef and pork. The demand for US chicken and pork is very high. With a total of 4 million tonnes of chicken meat and 1.7 million tonnes of pork, the US leads the global meat export business. The meat export market is expanding and the United States can meet the global demand.

Only 18% of the United States is fertile farmland. The remainder consists of deserts, semi-deserts, swamps, mountains and cold regions. On this valuable but limited area of land, food is grown to feed the world, even though the country is not obliged to provide the world with meat, nor should it offer meat for excessive consumption to its own population. If the USA were to drastically reduce the production of meat for domestic consumption and stop the unnecessary export of meat, it could become a pioneer in the worldwide cutback in meat production.

Expedient producers of meat

Several new states have taken over pristine lands which are partly suitable for agriculture. After burning and deforestation they began by raising cattle and subsequently, with the help of irrigation, they started to grow cash crops in the form of grain or oilseed. Initially the price of grain and oilseeds was very low, the transportation costs were very high and the markets were unreliable. The end of the East-West conflict, the gradual liberalization of China's economy and the subsequent globalization has brought new ideas into the field of agricultural products. The trade in grain and oilseed began to boom, and these now serve as raw materials for the production of meat.

Countries such as Argentina, Australia, Brazil and Canada have large tracts of land on which to grow cash crops and at the same time plenty of room for livestock. Instead of exporting all their grain and oilseeds, it was expedient to feed a portion of it to their own livestock. The results were astounding. With meat one could earn more than with cereals. Within a short time those countries began to produce large amounts of meat, which are competitively priced in the world market. Strangely, neither the price nor the demand have decreased. Instead, a new consumer class has gradually developed worldwide, which consumes meat produced in far-away countries.

In addition to the larger countries mentioned above, there are also smaller countries which produce large amounts of meat for the world market. Countries such as the Netherlands, Belgium and Germany buy animal feed on the international market and produce more than 100 kg of meat per capita. Two countries which stand out are Denmark and New Zealand. With the production of 430 kilograms of meat per head of the population, Denmark is the largest meat producer in the world. At the same time, the country produces about 1700 kg of grain per head of the Danish population, giving it the largest per capita grain production in the world (FAO, Statistical Databases, Rome 2015). So the small country of Denmark competes in fourth place (2011) after the USA, Germany and the Netherlands as an exporter of pork.

New Zealand produces little feed grain or oilseed, but the country has huge areas of pasture land where cattle and sheep can multiply unhindered. Small airplanes loaded with fertilizer and grass seed were used to sow pature in the barren wilderness, on which flocks of sheep could graze. These measures have made it possible for New Zealand to produce 443 kg (2012) of meat per capita, which has made the country into the second largest animal breeder in the world.

All the countries which produce meat surpluses have became export-dependent. The meat export market is flourishing, feed delivery can keep up with the demand and the purchasing power in the importing countries is increasing day by day. What are the drawbacks to this lucrative business? They only export the meat, but the slurry remains in the country, contaminating the ground water, polluting the air and spreading zoonoses. Countries could be compelled by law to export ten tonnes of slurry for every single ton of meat exported. However, exporting even the dry residue of the slurry will continue to pollute the air and at the same time affect the environment worldwide. Meat exporters monitor the quality of their meat, but ignore the destruction of the environment in their own countries.

Meat and the fate of dogs and cats

The most popular pets are dogs and cats. The dog has always been an omnivore. Leftovers, bones, and biological waste were his traditional food. In contrast, the cat[1] was never dependent on humans for their food. Until the second half of the 20th Century, the ordinary diet of a dog was leftovers and for the cat it was mice in the house or yard.

A slaughtered animal produces at least 30% waste, which can not be consumed as food. Industry has found new ways of processing this waste as pet food and they have become successful. Later, specially slaughtered meat and fresh fish, were also used as a dog and cat food. Industrial pet food made life easier for the pet owners and more and more cats and dogs were brought to the cities as pets.

Dogs and cats were provided with canned food and packaged dry food. During the First World War, canned food was introduced and mass-produced to feed soldiers. It was too expensive to use for pet food. It was only after the Second World War, as tin started to become cheaper to produce, that the canning industry expanded and began producing canned food for dogs and cats. Cats were provided with a tray as cat litter for nature's call. But as the dog does not pollute its own territory, the owner has to go with him for a walk. As a result, there is widespread dog waste in the streets, on the sidewalks and boardwalks, in pedestrian areas, on riversides and in the parks of industrialized countries.

---

1   The cat was never domesticated. They approached humans of their own accord. When man became settled, he had to store food, which in turn attracted rats and mice. The rodent population increased in ratio to food storage and finally the wildcat approached the settlements, because the cat had an interest in the mice. People saw the cat as an enemy of their enemy and supported its presence. So the cat became attached to humans, but was never given orders, because man never needed to tame the cat.

The market for pet food in developed countries is largely saturated. But globalization offers the opportunity for new outlets in the developing and emerging countries. Nouveaux riches adopt a largely Western lifestyle. Imported Western breeds are a growth market and some breeds have become status symbols. It is not hard to imagine who is behind the market in dog breeds. Firstly the dog breeders, followed by the canned pet food industry.

Much like their owners, dogs have given up their traditional methods of feeding and live largely an unnatural life. Dogs are neutered or sterilized at an early age, exposed to nutritional diseases and with increasing health problems they are ruthlessly given over to veterinary euthanasia. The fate of a modern dog is based on imprisonment, enslavement, unnatural food, sterilization, castration, isolation, eating disorders and euthanasia. Moreover, every year millions of people are bitten by those well-maintained dogs in industrialised countries. No other mammal species, whether wild or domestic, attacks human beings in this way. The number of injuries caused by dog-bites is highest among children aged 5 to 9 years of age. Every year, 4.7 million U.S. Americans are bitten by dogs and these dog-bites result in approximately 16 fatalities (CDC; Fact Sheet, Dog Bite, Atlanta, USA 2013). During the last five decades, dogs kept as pets in industrialised countries have killed and injured more people than any other wild or domestic mammalian. Street dogs of poorer countries are uncared for, they usually do not pollute their territory and they do not attack people, unless they are suffering from rare diseases like rabies.

Unavoidable contaminants in meat production

Meat was seldom contaminated in the traditional economy. Farm animals, such as dairy cows or draft animals, that had been slaughtered because of old age were strictly speaking not meat animals. Meat was a rarity, and it was tough and virtually free of pollutants. The goal of animal farming was not meat, but milk, eggs and muscle power. Despite higher prices, the meat was just a by-product. Often male animals on chicken farming were slaughtered prematurely, but these animals were not fattened beforehand, because feed was always a scarce commodity.

Modern man uses hardly any draft animals and he is not prepared to eat tough meat from an old animal. Man's nutrition has changed completely and animal products have become staple foods. Regular deliveries of milk, eggs and meat secure man's food supplies. These three basic products require labor-intensive and complicated conversion paths. Several complex industries such as feed production, animal breeding and meat production support this process.

The crucial difference between consumption and natural growth is that people have to eat several times a day and the plant food products whether produced in the wild or as a result of cultivation need several months to grow. For example, corn needs 3 to 5 months to grow before man can eat it. If a cow or a chicken is fed this corn in order to get eggs, milk and meat from the animal, the length of time until consumption is even longer. The period of time needed to produce food from animals is several times longer than the growing period of plant food. Producing cereal takes less than half a year, but the feed grain takes years to produce milk, eggs and meat.

An old saying from the tropics purports "Sudden hunger does not ripen the banana." Because the banana is harvested green, it can take between several days to several weeks for it to ripen naturally. But the modern banana industry has utilitarian science on their side. The bananas are transported in refrigerated containers across the oceans in their green state, and upon arrival are gassed in ripening rooms. This ripening gas, called ethylene (C_2H_4), is produced artificially and it is regarded as a narcotic, which is dangerous to health. But if people do not live in the area where the banana grows and still want to eat bananas regularly, they have to accept this unnatural assault. At room temperature, green bananas can ripen without a ripening gas. It just takes time and patience, which people unfortunately do not have and so they prefer the unhealthy option.

Growth stimulators, disinfectants, vaccines, antibiotics and many other chemicals are in use in modern meat production. A pig can live to be 24 years old, but on a modern farm it must weigh 100 kg to 150 kg at six months to deliver sufficient meat, or a chicken which can live to be 9 years old, is slaughtered after only 4 weeks because at that age it can produce enough meat for the industry. It is very naive to think or to demand "I have a right to clean meat". Science provides no alternative for producing unpolluted meat. Intermittent punishment of feed manufacturers, breeders or producers of meat products do not result in improvements; the problems will continue to exist elsewhere. The pollutants in animal breeding will continue to pass through the food chain to the consumer and modern medicine will not be able to help with this disaster.

Unavoidable outbreaks of animal epidemics

Animal diseases and epidemics are common natural phenomena which occur periodically. In modern factory farming these occurrences have a different effect than in nature or in traditional animal husbandry. At the outbreak of certain epidemics, all animals on factory farms could be affected and as a preventive measure a mass slaughter of all animals is carried out. Animal diseases in modern factory farming are numerous, but the most notorious of them are Swine fever, Avian flu, BSE and FMD.

Swine fever: Swine fever, also called hog cholera, is a highly contagious viral disease. Swine fever is omnipresent, spread all over the world and every year it destroys millions of breeding pigs. According to FAO this disease has never been brought under control. The incubation period of the disease can last from 2 days to 5 weeks and once the virus is in the blood, the symptoms of the disease become visible. The database on swine fever suggest that wild boars may harbour it and they are thought to be responsible for the spread of the disease. When there is an outbreak of swine fever in an area, slaughtering is the only means of preventing further spreading. In organized countries, there are strict requirements for reporting an outbreak of the disease and within a certain radius the animals are killed immediately.

Avian flu: This is a highly contagious viral influenza-like illness. It is called H5N1 in the fields of medicine and veterinary medicine. The highly pathogenic disease is also known as avian influenza. This pest-like disease is particularly widespread in southeast Asia among many species of birds and kills millions every year. But this disease spreads most fiercely in fowl and turkeys. According to the latest report of the World Health Organization this disease is out of control and can not be stopped. Since 2003, hundreds of millions of infected chickens and other breeding poultry, whether infected or not, have been destroyed worldwide every year. Because H5N1 cannot be fought, killing remains as the only protective and preventive means against this disease.

BSE: The cause of BSE (Bovine Spongiform Encephalopathy) is very controversial, but some sources maintain it could lie in the protein meal in cattle feed, which is made up of the carcasses of farm animals. BSE may be caused by certain proteins that transform the brain of cattle into a sponge-like mass with fatal consequences. The time from infection to the onset of this disease is 4 to 5 years, but the disease first becomes visible between the 4th and 5th years of age. However, all male calves and beef cattle are slaughtered in the first year of life. This means that if a newborn male calf has contracted BSE, it is difficult to detect it because the symptoms of the disease are apparent only much later. Only dairy cows are allowed to reach the ages of 4 or 5. When BSE was associated with CJD (Creutzfeldt-Jakob disease), it was taken very seriously around the world. According to the latest findings, both BSE and CJD can be detected using the same blood test. BSE has now spread worldwide. The killing and disposal of affected animals is so far the only method practiced to combat and prevent this disease.

FMD: FMD (Foot and Mouth Disease) is a contagious viral disease particularly in animals with cloven hooves, like sheep, goats, cattle and pigs. However, during an outbreak of FMD in a specific area, all herds, whether infected or not, are killed and disposed of. For example, in 2001 alone in the UK there was a total of 2030 FMD cases. As a means of controlling the disease and preventing its re-occurence, 6.5 million infected and healthy animals were killed. Among these animals were 4.9 million sheep, 700,000 cattle and 400,000 pigs. This caused a loss in the food industry to the value of 5.27 billion euros and 4.25 billion euros, which were paid as compensation by the government (DEFRA; Animal Health and Welfare, FDM Data Archive, London 2004).

Separate food markets

There was very little meat available in ancient times and the profession of butcher hardly existed. Due to the shortage of animal feed and the resulting lack of of domestic or farm breeding, very few animals were slaughtered. At the same time, due to the absence of suitable hunting weapons, wild animals were seldom killed for meat. There was no permanent employment as a butcher for this rare product from which a person could have earned his livelihood. Old or sick domestic animals were slaughtered by the farmer himself and animals which were hunted were also eviscerated by hunters themselves.

The modernization of the firearm between the last half of the 19th Century and the early 20th Century resulted in a new era in the history of hunting. Weapons used for hunting, which became increasingly more efficient, enabled wildlife to be killed in large numbers in a very short space of time. The hunter no longer had time to eviscerate or process individual animals. Because hunters were able to supply the population regularly with dead animals, butchers emerged as a new profession. The wild animals were very tough and butchers developed various methods to make the game more palatable and to preserve the meat. The word 'butcher' probably comes from the context of massacre, executioner and hunter. Indeed the word butcher means a person who processes or butchers animals which have been killed through hunting. In this sense a butcher was not a person who had to kill a living animal. Butchers and hunters worked closely together and butchers were always associated with hunting activities. Even the tough, old animals from the farms had to be butchered and minced.

Increased hunting activity and the availability of cheap weapons for hunting brought many wild animals almost to the point of extinction. Many professional hunters and many butchers became unemployed or looked for new employment. Those butchers who continued to work, took on the new job of slaughtering farm animals themselves. By the second half of the 20th Century, the butcher had become an established retailer, similar to a baker or pharmacist. There was no more venison and the supply from farms was irregular. Sometimes there was no meat in butcher's shop so people had to go without meat. The supply was abundant when animal epidemics broke out and the infected animals needed to be slaughtered.

The ultra-modernization of agriculture in the second half of the 20th Century allowed the production of large quantities of feed, which became a basis for intensive farming. The small butcher's shop could not compete with intensive animal farming. To carry out the mass slaughtering of animals, modern slaughterhouses were built. The same has happened in the case of fishing. When the motorized fishing industry took off, the smaller fishmongers could not handle the huge supplies from the ocean and so the massive fish catches had to be processed in factories. For reasons of cost, the products of the mass slaughterhouses and the fish factories were not offered at retail outlets, but in supermarkets along with other foods.

Cooling, dating, bright lighting, special packaging, etc. require numerous hygiene measures to be followed. These additional measures involve additional costs that must be covered by revenues from other commodities. The present fish and meat products also increase the risk of bacterial and viral threats to other foods such as fresh vegetables, fruit and baked goods, which are stored nearby. Contaminated meat from animal epidemics may jeopardize the health of people who had no intention of buying meat. Epidemics such as SARS, swine fever or bird flu are not predictable and uncovered meat increases the risk of the spread of such diseases. The same applies to fish and fish products, which are exposed to bacteria, viruses and parasites.

A separate fish and meat market, away from other foods, would reduce these expenses for the supermarkets, as well as reduce the price of normal food and minimizing health risks for consumers. Special hygiene measures only for fish and meat, sold in separate markets would reduce extra expenditure and consumers could buy these products from well-cooled and insulated markets.

Environmental and hygiene tax on fish and meat

Meat is not a primary food that is available directly after its production. Meat is a much-reduced accumulation of primary food products. Large amounts of feed and energy have to be invested to produce meat. Feed grain with a water content of 13% is used to produce meat which contains up to 80% water. Nevertheless, the same tax rate is levied on meat as on foods of plant origin. A product such as meat, that requires a lot of feed should be classified as a luxury-class merchandise and not as a staple food such as bread or potatoes. Accordingly, a new luxury tax on meat and meat products should be levied. This would reduce excessive meat consumption and thus promote health. It would not result in people with high purchasing power eating more meat than the lower income class. Delicious meals made with meat are no longer popular with the rich. Perhaps the new richly would consume an abundance of meat. Unfortunately they would have to live with the results a few years later.

The mass farming of animals requires extreme hygiene measures. The use of disinfectants on farms, in slaughterhouses, in meat processing plants and meat outlets causes huge environmental damage. The disinfectants, breeding chemicals, vaccines against animal epidemics etc. get into the food chain and affect humans. Over-exploitation of land to produce animal feed and related environmental destruction is directly connected to the consumption of meat. Animal excrement of farms pollute valuable fresh water resources, damage the air and reduce the standard of living. An eco-tax should be introduced to partially combat the environmental impact of the meat industry. Even so, money cannot repair the enormous damage wreaked by the consumption of meat. Contaminated groundwater cannot be cleaned, nor can the deforested jungles be returned to their original state. However, the levying of a luxury tax, a hygiene tax, a health tax and an environmental tax could reduce further damage to the natural world, including humans. There is an urgent need to remove agricultural subsidies for meat and fodder production.

Chapter III: The environment of meat production

A scavenger society serving the meat consumers

What is a scavenger? A scavenger is a person who has to do extremely dirty work for other people. From medieval times, scavengers had to dispose of the human excrement produced in the trading cities of Europe. Early in the morning, they carried the receptacle from the outhouses on their shoulder, full of human excrement and took it to a disposal point. Rural agricultural societies had hardly any toilets. They went into the fields or bushes to answer nature's call and this practice was also regarded as an act of fertilization. The new urban population began to empty their bowels at home and someone had to dispose of this waste. This inhuman practice went on until the early 20th Century. Running water, sewerage and electricity have saved these people from carrying out this horrible form of forced labor.

In intensive livestock breeding, animals are kept in small spaces and continuously fed large amounts of fodder. The regular feeding in factory farming causes the production of endless feces and urine. It is falsely claimed that vegetable proteins are converted to animal proteins such as milk, eggs and meat. In fact, over 90% of the plant feed is lost in the process of excretion and the so-called meat peasant or dairy farmer spends most of his life knee-deep in the feces and urine of his animals. A chicken produces on average about 100g of feces per day, a pig produces about 3 kg and cows over 20 kg (FAO/AFRIS). The farmer's sense of smell is dulled by the animal stink of feces and his body odor, even away from the farm, carries the smell of animal urine and feces. It is a tragedy that people must spend their lives in such filth and stench.

100 cows, the size of an average herd, produce 2000 kg cow dung and similar amounts of urine a day. Some farmers have hundreds of cows and the excrement, called slurry, is stored in large tanks. The farmer transports the slurry to the fields, sprays it on as fertilizer, creating a bestial stench in the air, he breathes in the methane-ammonia mixture and believes that this is the smell of good country air. Through his work, the slurry gets on his body. He firmly believes that this is the most natural form of labor input. Unfortunately he is wrong. There has never before been as much manure and urine on earth as now in industrial animal farming. The farmer has two things to deal with: feed and feces. The feed can have a pleasant smell, but feces and urine can in no way be regarded as pleasant and certainly not in the amount that the modern farmer has to contend with.

The so-called factory farmer must be in attendance on his animals almost around the clock, leaving only time for meals, sleeping, answering nature's call and for the management of basic tasks. He must milk the cows, collect the eggs, feed the animals and dispose of the slurry on a daily basis. He works 365 days a year, for life, sees no difference between Sunday and Monday, autumn, spring, Christmas or the New Year, because his animals must be fed and looked after. He is in daily contact with his animals, feeds and cleans them and finally when it is time to transport them to the slaughterhouse, he looks at them again makes eye contact and lies to himself "I have not deceived you". He may feel heavy-hearted, but tries to take it lightly. Supposing he is rich, he does not have time to enjoy his wealth because he is a slave to his livestock. This is the modern scavenger society working as forced labor in the service of meat, egg and milk consumers. If people want to eat meat and drink milk, they should breed the animals themselves, feed them, dispose of their excrement, milk and slaughter them themselves and should not force other people to do such inhuman work with their purchasing power. It is unethical and unjust to abuse people in this way. They must be compensated, relieved of this terrible task and receive a fair pension.

A comparison with the life of the arable farmer

Arable farming is as different from animal husbandry as flora is from fauna. It goes without saying that because plants do not have a central nervous system, they are not sensitive to pain. The life of a livestock farmer and that of an arable farmer is completely different. Misleadingly, both of these professions are called farmer. The term farmer is appropriate for those who cultivate land, but animals are not cultivated, they are kept under surveillance.

The size of the harvest is the most important factor for farmers. In fact, almost all harvests are conditioned to a particular season and labor input during harvest time is very intensive. Nevertheless, the workload depends on the particular crop. Different plants, depending on their species, can be harvested after a few weeks or after many years in the case of a permanent crop.

Cultivation, maintenance and harvesting are the three most important time-related factors in arable farming. The growth phase of plants, which is also called the maintenance phase, is the longest. However, the work involved in planting and harvesting is more intensive than during the maintenance or growth phase. For example, the cultivation of grain on a hectare of land needs one to three weeks and less than a week for its harvesting. A four-to five-month grain harvest has up to 80% idle time in which the farmer does not have much to do. The planting of potato seed on a hectare of land, including plowing, needs three weeks and harvesting takes only a few days. The potato plant grows in three to five months and the farmer has little to do apart from superficial monitoring.

With an industrial-type of cultivation of cereals or potatoes in a larger area, the farmer would have the identical time during the growth phase as during planting and harvesting. Plants must not be fed every day nor do they produce excrement or waste on a colossal scale. After the completion of a cereal or potato harvest, a large period of time is always available for other activities. The farmer does not stink of potatoes or grain, nor does he sink in crop residue. Of course the chemicals that are used to enhance growth and protect the plant are not healthy, but identical additives in the form of disinfectants, antibiotics or other protective products for animals are also detrimental to the health of an animal breeder. Whether he is involved in vegetable cultivation or has a fruit plantation, the farmer has time for himself and his family, he can enjoy life, he can participate in celebrations, sometimes he can sleep in and once the harvest has been brought in, he has a sense of satisfaction.

Meat self-sufficiency

Every human being has the right to determine his or her own food intake, but not at the cost of other people. Meat does not grow on trees but the breeding of livestock, slurry management and slaughtering are indispensable to the production of meat. No other food production requires as much effort as the production of meat. It would be better if meat consumers were willing to carry out these procedures themselves and produce the meat for their own consumption.

Modern man has a lot of experience in dealing with animals. He keeps dogs, cats, birds in cages and fish in an aquarium. These pets are uneconomical, unjustly imprisoned and neutered. At the same time they are fed and cared for, thereby enabling a form of bilateral slavery. People treat these animals with affection and ensure their well-being through medical treatment. They inform themselves about the habits of these animals, dispose of their feces and if they go on a trip, they will ensure that someone else feeds and takes care of these animals. So, the pet owner looks after his animals better than a meat breeder. A consumer of meat could breed or keep animals himself and thus enable a kind of subsistence economy. Because of the shortage of space in cities, smaller animals are bred. This kind of self-sufficiency could protect the environment, prevent the zoonoses, put an end to meat scandals, reduce fodder production, diminish the ethical concerns and abolish forced labor in the service of the meat industry. Here is a great alternative to commericially produced meat. The consumer would not be forced to eat the limited range of meat available today, but he could choose from a greater variety of meat.

The main advantage for the subsistence breeder is the quality of the products. The animals are supplied by the owner and therefore the final product is known to him. Moreover, this kind of self-sufficiency reduces excessive meat consumption and it is relatively healthier than eating meat bought from a supermarket. A major advantage of subsistence animal breeding is the disposal of biological waste and leftover food, as they can be eaten by the livestock. However, meat will persist as a very controversial food.

Meat producers - the bogeymen

Meat producers are not well respected in society. Even passionate meat eaters express peculiar criticism of meat producers. In fact the owners of the ultra-modern poultry farms and slaughterhouses are not spared from this discriminatory criticism. Despite their ostentatious, jet-set lifestyle and wealth, they are the losers in society. They are hardly visible in public life, nor are they asked to take part in charitable work or to take on political duties. Often they hide from the public, the press or from environmental organizations. They are blamed for factory farming, animal transport, animal cruelty and for the violation of ethical values. The meat producers must answer before the judiciary at every meat scandal or discrepancies in meat production. It does not end there. The owners of those facilities are themselves under constant mental pressure and the older they are, the more psychological stress they suffer.

The meat industrialists are also food traders. Other entrepreneurs such as the feed or dairy producers, operate on almost the same level, but they enjoy a higher status and they are not discriminated against by society, except in the case of active participation in a food scandal.

Meat consumers are in a majority in the population and they are regularly supplied with an unlimited amount of meat. Meat has even become a staple food. Why do the staple food producers have to hide from the public? When bread was a staple food, farmers were regarded as the backbone of a nation. However, since meat took over the role of a staple food, the meat producers have been defamed as the bogeymen in society. It is an ambivalent morality of meat consumers that they do not stand behind their meat producers. The meat consumer complains about meat prices, the quality of the meat, meat scandals, cruelty to animals, damage to health, environmental impact and do nothing to help the meat industry. It is a mystery why meat producers serve such ungrateful customers. Their answer is "If we do not do it, someone else will." But life is unique, and wasted time and opportunities cannot be recaptured.

The butcher and social discrimination

The butcher is a bourgeois profession, but the term "butcher" is used as a curse throughout the world. Even people who love to eat meat scold other people as "nasty butcher." They eat meat, but would be afraid to slaughter an animal. Their purchasing power is their way of getting out of this act of killing. This process is identical to the handing out of the death penalty, because the executioner has to carry out the hanging, but society regards this as the worst profession in the world. The employer, namely the judge, remains an elite upper-class personality. But the subordinate person, who has to carry out the sentence, is demoted to an untouchable in society. The death penalty would disappear if the same judge who pronounces the judgment, had to personally perform the execution.

World history is full of terms like slaughterer or slaughter. War, civil war, persecution or defamation are regarded as a kind of slaughter, and the perpetrators are known as butchers or slaughterers. In no language has it been possible to come up with another word for butcher or slaughterer. All previous attempts have failed because consumers of meat have refused to accept any other appellation.

Butchers are, in general, not rich people who can lead a life of extravagance. It is a normal job for them and they do not hide from the public. They perform their social obligations, their circle of friends is not limited, but people make strange statements like "I am not related to a butcher".

The glorious hunter

Many nobles, chiefs, princes, kings and emperors have gone in for hunting. Even very wealthy monarchs, who had no need to hunt for meat, have engaged in adventurous and perilous hunting activities. Whether a democrat or a socialist, a significant portion of the political elite still go hunting. They often shoot wild animals, which do not threaten them in any way. Despite cheap meat being available in the market, the wealthy elite often eat game meat, which smells of urine and tastes tough. These addictions, which are anchored at the genetic level, are regarded as sport. The simple definition of sport is - physical exercise for health or competition. The insidious shooting of innocent wild animals from a hiding place cannot be considered as promoting health, but as a primitive addiction to hunting.

Hunters in any era could claim they have a right to hunt, but the traditional right to hunt was based on food supply. Later, when firearms were modernized, the hunter described predators as being enemies of man and domestic animals. Later still, wild animals were hunted in the name of science and research.

Modern hunters justify their hobby as an activity which helps to conserve nature. They maintain that if the number of deer and stags weren't controlled through hunting, they would eat young trees, or damage their barks, or the population of wild boars would grow uncontrollably. They also claim that wild animals endanger road traffic and feed on food crops. However, the hunter contradicts himself because food production is largely in favor of livestock breeding and animal husbandry only became possible because hunters had wiped out the carnivorous species from the forests. Now, the hunter takes on the role of carnivore with a gun in his hand and does not see himself as a predator. If he encountered a predator like a wolf or a bear, he would recognize it as his rival. He would stalk the animal, kill it and stuff it as a trophy. The modern hunter feels like the God of the forest. Although he loves wildlife with body and soul, he thinks the law of nature allows him to decide when an animal must die. Indeed the hunter is addicted to game meat, but he does not want to admit it.

The population of moose and wolves in Sweden reveals how strange the idea of a modern hunter of wild animals is. There are over 400,000 moose but only 200 wolves living in Sweden. Nevertheless, the Swedish Environmental Protection Agency SEPA announced that 27 wolves could be hunted in mid-December 2010 in five major hunting areas. More than 12,000 hunters hastened to the chase, shot most of the permitted number of wolves on the first day and injured a large number of other wolves (WWF, The wolf hunt in Sweden 2010 and 2011, Stockholm / Solna 2012).

The main diet of wolves in Sweden consists of moose, which they hunt in packs. Now the hunter has assumed this role, so that moose hunting has become one of the most popular sports in Sweden. The hunters in Sweden kill more than 100,000 moose a year, an amount which could feed 50,000 wolves. The following text shows how strange the opinion of Swedish hunters is regarding their native wolves:

"The wolf is a large predator, and generally able to kill a man. The last incident of this kind happened in 1820 in Sweden, and it involved a half-tamed wolf in the province Gästrikland."
(Brunnvalla / Sweden - Land und Leute / Wölfe, 2012).

A half-tamed wolf had killed its owner almost 200 years ago, but little is known about how many of those 250,000 hunters were killed during the annual autumn moose hunt in Sweden.

What distinguishes a hunter from a poacher? The simple answer is that the hunter is licensed and the poacher is not. Poaching arose in order to fight hunger and for financial gain. But the modern hunter is obsessed with shooting innocent wild animals and the consumption of wild meat is not due to hunger, but a genetically-anchored addiction to game meat. Breeders, butchers, meat traders and meat industrialists work to meet the supply, but a hunter kills wild animals as a hobby and yet he is well-respected in society. It is easy to understand where this recognition comes from, because it is the social elite who engage in this hobby.

Veterinary doctor – a childhood dream becomes a nightmare

Children's dreams about what job they might do one day are numerous: pilot, captain, firefighter, engine driver, veterinarian, footballer, singer, actor, truck driver, excavator operator, bus driver or ice cream seller are all wishes expressed by children. All of these dreams can come true, apart from the profession of a veterinarian. Children want to be vet, because they want to help animals. Children want to give everything so that animals can live in a better world. But this dream job of childhood turns into a nightmare. Instead of helping animals, many of those childhood dreamers end up in slaughterhouses as meat inspectors. They organize mass killings during animal epidemics, they perform immoral castrations and sterilizations or cull sick animals. No child aware of this reality would choose a career as a veterinarian.

It is not necessary for veterinarians to walk through the ranks of halves of beef or pork, nor do they have to perform artificial insemination. Wild animals certainly do not want to be treated by man and man has no responsibility for them, unless they are directly responsible, as in a tanker accident or poisoning by pesticides. The commercialization of veterinary medicine is carried out directly through industrial animal breeding and in relationship with the mass consumption of meat. In this case, the modern veterinarian is more of a meat, egg and milk doctor who has little to do with animal welfare. No science is needed to kill a newborn calf or chick shortly after birth just because they are male. Modern veterinary medicine knows no ethics and morals. Profit is their primary objective and euthanasia is their best weapon.

A minority of conscientious veterinarians try as far as possible to help nature. They are actively involved in environmental organizations, try on their own initiative to assist stranded marine mammals by taking them back to the open sea, they struggle to free captive sharks from drift nets or they actively demonstrate against animal testing. But they suffer from a chronic lack of financial resources and they are often dependent on donations from the public.

The first step for a veterinarian to return to reality would be to give up the consumption of meat. The animal world is not limited to cattle, pigs, poultry and pets, but there is an endless variety in the animal kingdom. In reality, a veterinarian is a wonderful, interesting and secular profession. A veterinarian does not need to put himself at the disposal of the meat industry. As a recognized academic, he or she has many other opportunities to earn livelihood and at the same time help endangered animals. The childhood dream of being a veterinarian could be realized if he would only regard animals as part of nature and not as useful objects.

The patchwork of the Earth

The landmasses on earth are green. The plant world, without natural enemies, would have taken over our planet to the point of suffocation, if it was not kept under control by numerous herbivores. They eat the green plants and allow the propagation and airing of the woods. Adequate food means population growth. Herbivores would multiply, eat the vegetation bare and convert the green planet into a brown planet if predators did not keep a check on the number of herbivores.

There has been no rapid growth among the larger predators. The causes are difficult hunting methods, laborious feeding practices, a short lifespan and early death from diseases, rivalries and hunting injuries. In general, herbivores live longer than meat eaters. Rivalries and feeding injuries among herbivores are very rare. Herbivores do not compete with other herbivores for food. They do not practice infanticide, the killing of their own or another animal's offspring, which is commonly found among carnivorous animals.

Symbiosis between plants and animals means that the ground is fertilized and kept fruitful. Fertile soil retains moisture and allows the growth of plant and animals. The forests created by humans, do not fulfill natural requirements and at the same time destroy the balance between flora and fauna.

The latest satellite images reveal clear evidence that the settled areas of the earth are completely without forests. The visible individual forest patches are planted forests where joggers run and families go for summer picnics but there is no diversity of wildlife. From an altitude of 8,000 meters, even with the naked eye, the green landscape of the earth appears as a geometric patchwork (book cover), ranging in colour from gray, to light green, dark green, yellow and dark yellow. The colors indicate the stages of cultivation of the arable land, from land that has just been plowed, land with new germination, fields with growing crops, harvesting and land that has just been harvested. This artificially-coloured earth is in sharp contrast to the deserts, where, apart from desert herbs, no other plant can grow. Apart from in the mountain and desert regions, the warm and temperate zones of the earth resembles a giant patchwork in different shades of yellow and green. This dangerous camouflage represents an impending threat which must be taken seriously. The patchwork of the earth is a symptom that the symbiosis between plants and animals has been destroyed.

Animal feces and soil fertility

The earth does not have enough fertile soil. The land which is fertile makes up an estimated one-tenth of the available land area. This scarce arable land is increasingly threatened by animal feces of industrial livestock, called slurry. The greater the production of meat, the higher the quantity of slurry, which is poured onto fertile arable land. A small volume of manure can be used as fertilizer but constant use of slurry destroys the fertility of the soil and in addition contaminates the ground water. All industrialized nations with industrial meat, dairy and egg production have suffered for a long time from the stress of the daily volume of slurry. The phenomenon of slurry management is increasingly a global problem. With the industrial nations as a model, the remaining countries of the world are competitively producing food of animal origin.

The first goal of developing countries is the acquisition of sufficient animal feed. This is a huge undertaking and they are trying to produce animal feed from different sources. The local meat industries are involved in trying to establish this lucrative business. Once the problem of feed shortage has been overcome, developing countries will go in for meat production. The slurry produced would at first be regarded as valuable fertilizer, but gradually it would become a disaster, as in the industrialized countries. Deserts used for dumping slurry are too far away, and it would be too expensive to transport animal feces for thousands of miles to dispose of them. Slurry would be disposed of there where there is fertile land, food is grown and livestock is bred.

Proponents of factory farming maintain that the volume of slurry can be used as a valuable resource, as a fertilizer for organic farming. This argument falls short, since human defecation which produces more valuable excrement than others is disposed of almost unused. Human feces have higher levels of nitrogen, carbon, potassium, calcium and phosphorus, which could be a useful product if recycled and it would be better than using chemical fertilizers. The daily feces produced by a world population of over 7 billion could replace all the fertilizer factories in the world. But modern man answers natures´s call in an enclosed cubicle, too ashamed to take this step, preferring chemical fertilizers that are gradually destroying the soil and leaving nothing behind for future generations.

It is no mere hypothesis that excessive animal feces destroy soil fertility; it is a statement of fact based on empirical findings. Numerous small islands in the oceans consist of bird droppings. The bird droppings, called guano, are a valuable fertilizer if distributed in small amounts on less fertile land. But a guano island itself remains barren because plants cannot tolerate such a high concentration of animal feces.

In countries with intensive animal husbandry, the fertility of the soil will gradually be eroded. This will result in reduced feed production and there might be a threat to arable farming. In a few decades, the world population will grow to over 10 billion people. According to recent economic growth, they are expected to consume 100 kg of meat per capita. In addition, there is a demand for milk and eggs. An unimaginable amount of animal feces is expected in future generations and they will undoubtedly discover that feed does not only mean meat, but also slurry and by then most of the fertile parts of the world will have been become guano islands.

Meat and the world of leather products

The leather industry has developed in the same way as the steel industry. The workshop of a blacksmith, a one-man operation, has become a massive modern steelwork. The leather industry is one of the most important manufacturing industries in the world. Shoes, jackets, suitcases, bags, belts, furniture, purses and a wide range of other products are made of leather. The leather industry is a major employer and it is also an important growth sector. The future of leather is very promising because leather is a natural product which is very durable and at the same time it is considered stylish. The most important factor behind its increasing popularity is the falling price of leather goods. People who previously wore nothing on their feet, now own several pairs of leather shoes, and the traditional leather jacket which was handed down from father to son, is now available to all income groups and is a sign of prosperity.

The drop in the price of leather goods is due to the increase in leather production. About 90% of the world's raw leather comes from cattle and nearly 10% from sheep and goats. Cowhide is a large piece of skin, very durable, easy to process and very cheap to obtain. The dairy industry and beef industry are two independent branches that supply leather as a raw material. In fact, the cattle industry has three separate branches, namely the production of beef, milk and cowhide. Each sector claims it is the most important. When the production of milk in dairy cows drops, the cows are processed into meat, and sooner or later the male calves are slaughtered; in both cases leather can be produced. The younger the calf, the softer the leather. Renowned watch-makers or purse manufacturers advertise their products as "made from the finest calf leather." Cowhide has established a market of its own and, regardless of the fluctuations in milk and beef production, the importance of this product will remain. The worldwide demand for cow's milk is enormous, so that the supply of leather in future is guaranteed.

A question that needs to answered is: how important is leather to human life? Using animal skin to produce consumer products is not important anywhere. Leather was unimportant in ancient times. Animal skins were worn in the colder regions but were not common in other regions. Leather gradually became popular just after industrial salt began to be produced, because salt was used in leather processing. At the present time, there are abundant products which could be used as a substitute for leather. Plant fibers of all types and modified raw materials of plant origin are excellent leather substitutes. In addition, organic minerals such as petrochemical substances perform better than natural leather. It is not only unethical to enslave animals for their skin, but it is a sin in the vast environmental sense. Wearing leather is a status symbol and giving up this practice is an individual choice.

Luxury leather products such as leather furniture or clothing should be avoided. There are numerous synthetic products or products of plant origin that could replace leather. The meat industry calculates the value of leather as part of its total revenue. If for some reason the consumption of meat dropped, animals would continue to be bred for the purposes of leather production. The meat would then be used as an industrial raw material or as animal feed.

Meat and the bottled water industry

Although the Earth could be re-named the water planet, clean drinking water has always been a scarce resource. Contaminated water causes diseases. The causes of water contamination are mainly chemical and biological. Toxic chemicals such as arsenic, fluoride, artificial fertilizers or pathogenic microbes such as bacteria, viruses or parasites can contaminate drinking water. Knowledge about drinking water has increased enormously over the past fifty years and it came to be regarded it as a pure commodity. A global drinking water industry, which sells potable water in plastic bottles, has sprung up in a few decades. Water and air are not foodstuffs but life-sustaining elements, and one of them has been turned into a merchandise.

Fresh water is available in abundance. Rain alone provides masses of clean water. Less than 1% of the amount of rainfall would be sufficient to supply the entire world population with fresh water. More than 80% of the world's population lives in the rain belt where the bottled water industry is successfully earning money. Fresh water is promoted as the oil or gold of the 21st Century.

The contamination of drinking water is largely due to industrial agriculture and animal husbandry. Agro-chemicals and slurry contaminate the ground water and other surface inland waters. The bottled water industry is profiting from this development and people think of themselves as modern and civilized. The transportation of drinking water, along with the consumption of energy in the form of mineral oil cause some of the worst pollution because water is the most natural and essential product and the unnecessary use of fuel for these purposes can be regarded as pollution.

The production of meat, eggs and dairy produce claims over half of the world's arable land. Fertile land is cultivated for feed grain and artificial fertilizers or pesticides are used to support meat production rather than growing food. The global picture of the supermarkets with a large variety of bottled drinking water is impressive, but the growing mountains of empty bottles is devastating to the environment. For how much longer can the industry fill bottles with clean drinking water, if agricultural chemicals are seeping deeper and deeper into the ground and contaminating the fresh water? Meat production contaminates drinking water and the consumer does not see the relationship, although he and his descendants are directly affected by it. It is no superstition that animals who are enslaved unjustly and slaughtered, punish mankind in this way.

Meat and fertile land

The Water Planet Earth has little fertile and habitable land. Approximately 10% of the surface area consists of arable land. The remaining 90% of the land mass is made up of desert, semi-desert, cold regions, mountains, swamps and dense forests. Since the beginning of agriculture, the need for fertile land has become greater and greater. The availability of arable land has not been able to keep pace with population growth and has stagnated since the middle of the last century. Indeed it has only been possible to maintain food supply by practicing intensive farming. All possible measures have been taken in order to gain new farmland. Land reclamation through further deforestation and by irrigation in the desert are on-going practices. Deforestation causes forests to dwindle and the irrigation of the desert causes the desiccation of rare rivers in arid areas. The groundwater level will decrease and the great dry land trees will cease to exist.

The growing need for agricultural land is not primarily for the cultivation of food crops, but for fodder production to feed meat-producing animals. The feed industry is one of the most powerful large-scale industries in the world. It determines the supply of dairy products, eggs and meat as part of the modern diet. If human beings largely avoided eating food of animal origin, at least half of the arable land would automatically revert back to its original form of wilderness.

Enmity between man and predators

Agriculture and animal husbandry did not develop in tandem. The first agricultural activities were carried out on naturally-growing food plants such as bananas, coconuts, dates, yams, taro or wild cereals. These plants were partially protected and maintained by humans. In order to grow certain targeted crops, man began to till the soil. Tillage, especially for a single crop such as grain, needed a lot of muscle power. Using tamed animals to labor in the fields began with the introduction of cereal farming. However, their number was relatively small. Trees producing crops such as dates, coconuts, bananas, walnuts or other fruit need no special soil preparation done by the muscle power of tamed animals.

The disruption of these activities came with the cultivation of grain, for which the muscle power of animals became necessary. Later, more and more additional uses for domesticated animals were discovered. Animal breeding grew as a result of an increase in the demand for milk and meat. The herds foraged for food in nature. The beginning of animal breeding led to food rivalry between humans and predators. Predators such as the many species of cat, wolves and bears attacked domestic animals which were bred for their milk and meat. Only those working animals which were accompanied by people were rarely attacked by predators. An eternal enmity between man and predators had begun. The instinct-oriented and genetically-anchored instinct of predators is to kill herbivores, thereby keeping a check on their numbers. This instinct drives predators to watch the herbivores, regardless of whether they live in the wild or under human protection.

In the past, people lived in villages and settlements, practiced agriculture and feared the dangers of the wilderness. Traditional weapons for offensive or defensive purposes against dangerous predators were inadequate. Man needed a weapon with which they could defend themselves from a safe distance and fight effectively against those enemies. They could defend themselves during daylight hours with traditional throwing and thrusting weapons, but in the dark, these weapons were useless. The only alternative at night, when the predators were active, was fire. Traditional peasants drove predators from settlements or protected themselves against an attack on their families and domesticated animals, with the help of fire. Clearing forests with fire served not only an agricultural purpose, but was also an attempt to drive predators away. It was known that predators are afraid of fire and try to avoid coming into contact with it. To drive predators away, flaming arrows were shot from a bow and burning torches were thrown at them or a fire wall was lit as protection. However, a better method was sought with which predators could be pelted with fire from a safe distance. With the invention of gunpowder, a weapon was invented with which man could attack animals from a safe distance. After the invention of the firearm, the dangers from the wilderness gradually lessened and instead of hunting on the edge or outside the forest, people occasionally chased wild animals into their forest habitat.

Predators gained supremacy only when man did not have sufficient arms to defend himself. However, man gradually mastered his dominion over the animal world. The modernization of the firearm facilitated the expansion of animal husbandry. The predators were eradicated and the territories in the wilderness were declared danger-free zones. Predators could no longer chase domestic animals and so, with no natural enemies, their population grew. The forests were cleared to win farmland and pasture, and at the same time the remaining predators were hunted. Predators, which once dominated the forests, survived only in the circus cage, in zoo-enclosures, in the so-called National Parks. If an intruding predator like a wolf or a bear randomly attacked a domestic animal, they were mercilessly hunted and killed.

The forests have disappeared and the landscape has been shaped by human hands. Livestock and fodder production, as well as the maintenance of pastures have become the three most important activities on fertile land. Predators and the forests disappeared, because man began to regard meat as a staple food. Man took on the role of predator when he began to eat herbivores. The change of role might have been acceptable if man had hunted the animals in the wilderness, eaten them raw and lived in the forests. But this was not the case. People live in high-rises in the cities, they send animals to slaughter in factories, where they are dismembered and crushed, they treat the meat with heat and then eat it.

What is missing in nature? Numerous plant and animal species, which were unknown to mankind, have been lost. Wild animals eat wild plants that have never been cultivated as a crop. All those plant species have been lost through the expansion of agriculture. Furthermore, animals which were not important for meat production were not bred. They also disappeared from the forests. Nothing is known about the existing symbiosis between those plants and animals. The million-year-old balance of nature was destroyed in less than half a century. With no further thought, but in the name of growth and prosperity, humans justify the on-going devastation.

The leopard phenomenon

After lions and tigers, the leopard is the third-biggest of the big cats. The various species of leopard, such as the snow leopard, jaguar or panther have spread from Africa, Asia and Northern Siberia to the continent of America. In areas where all three are present, the leopard ranks in second or third place after the lion and tiger. This means that the lions and tigers keep the populations of the leopard down. This behavior is due to the competition for food which is why lions and tigers kill leopard cubs. Wherever the lion or the tiger dominated the forests, the number of leopards was very low. Tigers and lions were seen as the greatest enemies of leopards in the wilderness. But from the mid-19thCentury these two rulers of the wilderness were systematically exterminated. The leopard was not on the list of enemies of man because it did not steal domestic animals from the stable or the pasture, nor was the leopard a great risk to human life. The subsequent lack of natural enemies enabled the rapid growth of the leopard population. As a result of their increased numbers, they try to break into the settlements in search of food and they attack small animals like pigs, goats or sheep thereby entering into a new enmity with man.

Again, modern firearms were used against leopards and the International Union for Conservation of Nature and Natural Resources (IUCN) has registered the leopard on the Red List of endangered species. So, the leopard is a protected species, but the IUCN has made no further recommendations about the possible food sources for of this endangered cat. The salvation for the leopard would be the return of lions and tigers to the forests. In this situation, the leopard would stay away from man. But which solution is suitable for the big cats if meat remains a staple food for modern man?

Alone against nature

Before human beings became civilized, there was little difference between man and animal. Man lived in nature and reached a similar age to other mammals of a similar body size. Man was able to break his direct dependency on the wilderness and he is now in an endless struggle with nature. Man fought against the predators who had stolen their livestock, against the flocks of birds that had eaten their grain, or they fought against plagues of insects. These kinds of conflicts erupted only because man had produced something edible for his own survival and thought it was his own property. The uninvited guests out of the wilderness became enemies and so began the clash between man and nature.

Human beings have become adept at artificial living and have eliminated almost all major visible enemies. The remaining enemies are small and invisible species. The smaller and invisible creatures also multiply in proportion to the availability of food. A larger population requires a greater production of food and at the same time a large number of pests develop alongside. Pests are far tougher than the dangerous animals out in the wild. Conventional weapons like guns are useless and so chemical weapons are used. Protective chemicals are indispensable weapons in agriculture, in food production, food processing, food storage and consumption. These protective chemicals are used on land, in water, in the air and also in the production of food. Man thinks that he has been successful. This is pure self-deception since their own health and their own environment are in danger.

Modern man has chosen only a few of the innumerable plants and animals which could be useful to him. There was no space left for these others and so they were mostly destroyed. Some animal and plant species that are found in zoo enclosures or within a botanical garden have little importance for nature. Man's primary goal is to cultivate only those crops which can be used as feed for animals or food for humans. Profit-oriented science has largely destroyed a wide range of flora and fauna. Scientists would never admit it, but during the same period, they have created huge numbers of invisible opponents. These are the micro-organisms that are increasingly forming a formidable front against humanity. The extent of these micro-organisms is completely unknown but they are ubiquitous. Thus, man is increasingly alone against nature.

The so-called UN Climate Conference

The origin of the UN Climate Change Conference, which is also known as the Kyoto Protocol, goes back three decades. With the goal of reducing greenhouse gases, world communities are acting together under the leadership of the UN. Over 10,000 participants from three different groups, the delegates, observers and media representatives, gather for this climate conference. It is one of the most prestigious global conferences on climate change, which produces a lot of publications but achieves nothing for the environment.

The primary goal of this conference is to get industrial countries to reduce the emission of climate damaging greenhouse gases such as carbon dioxide, methane and nitrous oxide. Many discussions, lectures, reviews and agreements take place, but they do not deal with the main cause of climate change. Approximately 72% of greenhouse gases consist of carbon dioxide, 18% methane gas and 9% nitrous oxide. The majority of carbon dioxide emissions come from industry and transportation, whereas methane and nitrous oxide come from agriculture, mainly from animal husbandry.

The more meat is produced, the more carbon dioxide emissions there are. According to theoretical calculations, the production of one kilogram of beef causes up to 36 kilograms of carbon dioxide. On the basis of this knowledge, the idea of reducing meat consumption is being promoted. This idea was received with confusion because meat comes from animal carcasses and meat animals are fed with fodder of plant origin and in some cases with the addition of fish meal and carcass meal. What does this have to do with carbon dioxide affecting the climate? This is the common reaction of active meat consumers.

The most important goods transported are food-related raw materials and finished products. Food products are continuously sold and consumed, because these are the only energy source for the life of a seven-billion-strong world population. Other consumer goods such as clothing, medicines, electronics and iron ore are of secondary importance. Even the players of the non-food industries are directly dependent on food supply. In plain language food is the fuel of life. Because modern man does not cater for himself, food must be transported for his continuous consumption. Here, food of animal origin makes up most of the tonnage and travels longest distances. In the production of agricultural chemicals, industries need to emit significant amounts of carbon dioxide, or a large amount of electricity is needed for cooling and storing animal products, which in turn create carbon dioxide emissions.

All countries that produce large amounts of meat per capita also produce most of the greenhouse gases; the less meat, the less greenhouse gases. The best example of such a country is India. India produces only one ton of greenhouse gases and 5 kg meat per capita per year. All other countries that produce more than 50 kg of meat per capita and produce ten times as much greenhouse gases (United Nations Framework Convention on Climate Change, GHG data, Bonn 2012). India is an industrial as well as an agricultural country. But the country is not focused on meat production, because most Indians have plant-based diet.

The UN Climate Change Conference would have the solution to the problem of greenhouse gases, if they carefully examined the impact of modern meat consumption. But their hands are tied because at the same time the UN supports and promotes fodder production, animal husbandry, dairy farming and meat production with full vigor. This hypocritical climate conference is actually damaging because it refuses to recognize the core problem, namely destructive meat-oriented agricultural production, so it continues to waste public money.

The most important chapter in projects in trade and industry is stakeholder management. Stakeholders are those who have an interest in a particular project. These interested parties can be proponents or opponents of the project. Here the supporters are almost completely ignored and the opponents are examined carefully. The opponents are accorded a great deal of attention and all possible effort is made to bring them on board, or they are prevented from interfering with distractions. The diversionary maneuvers are of great importance, in that the interest of the opponents can be channelled somewhere else. The topics of the diversionary maneuvers are often very abstract, with no beginning and no end. For example, the best campaign shortly before a general election claims that the country is threatened with an invasion and the government calls for national unity.

The UN climate change conference is an example of true stakeholder management, in favor of global economic growth. The term climate is abstract, as abstract as faith, love, emotion or world outlook. Climate has no beginning, no end and no limits. Climate consists of innumerable and invisible components. The total volume of the atmosphere is many times larger than the earth itself and a hand-full of amateur scientists[2] try to force a change. This is not a fight between David and Goliath, it is pure self-deception.

Utilitarianism and academic negligence

The education system provides basic school knowledge and new generations specialize in particular tasks. In technical and scientific fields, they become dental technicians, surgeons, aeronautical engineers or food chemists. In market economies, qualifications are valued along with productivity. Finally, these kinds of technicians or scientists are specialists in a competitive field to offer a better product or to drive consumption

---

2 Amateur scientists are those scientists who are not supported by trade and industry for profit maximization. They suffer from financial misery and accordingly their possibilities are limited.

propaganda. They are specialists in their own area, work for a life time and then retire.

Disciplines such as sociology, social sciences or philosophy have a lot of freedom to explore society, to evaluate it and make recommendations. But most of these scientists keep themselves busy with medieval worldview theories, theories of social actions, structural changes, rationality theories and use complicated words that are not understood by the layman. Many of them seek collegial recognition, thereby forgetting their social obligations. The humanities have no major projects, they deal with an abstract synergy of knowledge. The established academic humanities are equipped with many packages which have very little content. Intentionally or unintentionally, euphemism is in many ways the most important element in the field of humanities. The many works, creations or essays of glorious scholars line library shelves like the bricks in a wall but hardly anyone understands or read those writings.

Chapter IV: Relinquishment of meat consumption

Meat and cultural conflicts

Modern world culture is largely dependent on faith. Religious influences determine the eating habits of a society. Christianity, Judaism, Islam, Hinduism, Buddhism and Jainism, all these great religions, influence the eating habits of more than 80% of the world's population. All these religions have their own dietary codexes which relate predominantly to food of animal origin. Jainism allows no meat to be consumed. Buddhism forbids the killing of animals and the use of weapons against other living beings. Meat consumption in Hinduism is partially permitted, but the slaughter of cows and the consumption of beef is absolutely taboo.

In the Abrahamic faiths, major differences prevail regarding particular types of animal and how they are to be slaughtered. In the Old Testament, only ruminants with cloven hooves were approved for human consumption (Leviticus 11). Animals such as pigs do have cloven hooves, but they are not ruminants. On the other hand animals like horses are ruminant, but do not have cloven hooves and according to the Old Testament the consumption of pig and horse meat is not allowed .

However, in modern Christianity the prohibition of pork and horse meat has been largely abolished. But pigs in Islam and Judaism are regarded as unclean and untouchable. At the same time, there are strict slaughter regulations in both of these faiths. The religious dietary codexes focus mainly on the consumption of meat and rarely on foodstuffs of plant origin. Meat as food is important in Abrahamic faiths, but meat in other world religions such as Hinduism, Jainism and Buddhism is partially or completely banned.

The intake of food in daily life is of primary importance and all other activities, such as working, sleeping or answering nature's call, are directly dependent on food consumption. All other activities are not feasible without food. The quality of a meal is dependent on the quantitiy of meat. The more meat, the more affluent the society. The choice of meat separates one society from another. The presence of meat or the type of meat determines which cultural group can participate in a particular meal and which not. Eating together is the first step towards mutual understanding and a peaceful communal life, but it can happen that because of meat, this collective step towards peace is prevented. A new hope for reconciliation and friendship could be established, if people abstained from serving meat at all intercultural events.

Carnivorous species do not lead a community life in nature. They live singly or in their own herds and do not tolerate other carnivorous animals in their territories. To avoid competition for food, feline animals like male lions, tigers or leopards kill their own young, and lions and tigers kill leopard cubs. In contrast, among herbivores there is no conflict about food as they eat the same food together at the same time. The initial clashes between different groups of human beings were based predominantly on food. This food was certainly of animal origin, such as prey from hunting or prey which was stolen from a leopard. As long as humans practiced no agriculture, there was little conflict over the fruit of a tree, because the gatherers moved on. The disputes over hunting grounds and the animals which were hunted are one of the oldest conflicts between clans, nobles or hunters. For example, if a wounded animal ran from one hunting ground to another, the hunters pursued it or demanded its release, causing a conflict. This kind of hunting often developed into a military conflict. As man began to breed animals, cattle rustling led to the most widespread conflicts between ranchers and rustlers. As the breeder increased his herd, the rustler would take advantage of it.

Man began to arm himself not only to hunt, but also for self-defence and to attack. Offensive and defensive weapons soon outnumbered hunting weapons. The original armament of mankind was based on meat procurement and not to enforce social ideas through force of arms.

People who do not eat meat, rarely possess weapons. The traditional meatless faith communities have handed over collective defense to the authorities. On the other hand, the private owners of arms are usually meat-eaters. Perhaps this behavior is a basic instinct triggered by the foreign meat in the body, which sounds an alarm for self-defence. Increasing global armament is a direct result of rising global meat consumption. Peace is inconceivable if meat remains as staple food, as important food or as a dietary supplement.

Milk-related cultural conflicts

The Hindu religious communities of India regard the cow as sacred. The reasons for this are easy to understand. The cow is used in agriculture as a draft animal, it is a supplier of milk and produces cow-dung, which is used as fertilizer. This precious animal is highly revered in Hindu society and cattle in Hinduism is regarded very highly. Despite these enormous advantages of keeping cattle, there are also numerous inconveniences associated with it. For example the cattle on the Indian subcontinent have brought the peasants into conflict with local tigers.

The spread of British colonial power was not a welcome development for the Indian farming community. The colonial administration was based on land tax, which caused suffering among the farmers. Land and cattle were the farmers' only two possessions. The high tax on land levied by the colonial government brought smallholders to their knees, while tigers stole cattle out of barns and fields. Then, the Englishman came with his new gun and hunted the tiger. The tiger population declined and the cattle population was no longer in jeopardy. Agriculture expanded, the revenue from the land reached a new record high each year and the colonial administration became more and more stable. Only because they hunted tigers, were Englishmen then welcome guests in rural India. That the Englishman also ate beef was not considered a serious problem.

By the end of the 19th Century, the gun was nothing more than an old musket. They had a short range, were inaccurate and inefficient. Until that time, hunting tigers in India was not very successful and the tiger still dominated the forests of India. When in 1887, ballistite or smokeless gunpowder was invented, it revolutionized the production of accurate firearms with a wider range. It took only a few years before this efficient weapon was used for hunting tigers in India's jungles. Many exciting hunting stories were published and in the first four decades of the 20th Century, India became a hunter's paradise. People with purchasing power from all over the world traveled to India to hunt tigers. Kings, nobles, officials, police officers, traders and planters all joined adventurous tiger hunts. From the north to the south, from the east to the west, in less than 50 years, tigers had been hunted across the country and systematically exterminated. Until the beginning of the Second World War, such glorious hunting stories were published widely in all major world languages. Many of those heroes of organized tiger hunts were regarded as life savers and the tiger went down in history as a brutal man-eater.

Lord Curzon, the British Viceroy of India (1899-1905), organized tiger hunts with 200 to 400 elephants and once killed 28 tigers on a 5-day hunt (Kauffmann, Oscar; Aus Indiens Dschungeln, S. 159, Bonn/Leipzig 1923). This merciless killing of tigers became the curse of tiger, as can be seen in Indian history.

The modern weapon has helped man to triumph over the tiger population of India. Regardless of the damage to the environment, this threat from the wilderness was defused.

Deforestation, the expansion of land planted with crops, the growth of the human population and the increase in the number of cattle were the result of the systematic elimination of the tiger. Land revenues and grain production increased and India became the largest and richest colony of the British Empire.

The tiger has disappeared, but a new social conflict has emerged. India has the largest cattle population in the world (currently over 330 million animals), and the majority of the population are Hindus, who breed cattle. If a cow is old, it can no longer work in agriculture, it can no longer produce calves or milk. This worn-out animal is then sold in a cattle market. The buyers of these old animals are not the Hindus, but members of the Muslim, Christian and other non-Hindu faiths. People buy an old cow not as a draft animal or in the hope that it will still produce milk, but to butcher it and consume the meat. Cattle slaughter is not tolerated by the Hindu majority and for this reason numerous Hindu-Muslim conflicts began, which became a part of everyday life in India. Apart from the slaughter of cattle, there are hardly any other social points of conflict between Hindus and Muslims in India and this situation has arisen based solely on the absence of the tiger. As a result, many Muslims demanded a separate state and the country was divided into two parts, India and Pakistan.

Before the colonial period, there was very little cattle and beef did not feature in the usual diet of Indian people. Regardless of their faith, cattle was valuable to all farmers. To Hindu communities, they were not only valuable but also holy. Once colonial rule was established, the tiger disappeared. The largest cattle population on earth, which is to be found on Indian soil, may not reach their natural age because maintenance costs are very high and there is a lack of space and fodder. But old, inefficient and unwanted animals may not be slaughtered. Because of the differing value accorded to cattle, Indian society dissolved into a conflict between Hindus and Muslims.

Cows are sacred to Hindus and should therefore be treated respectfully by their owners until their death and should not be sold to a butcher. A law should be introduced forcing cattle breeders to care for their animals up to the end of their lives and the sale of cattle should be forbidden because it is unethical. The unpleasant and harmful Hindu-Muslim conflicts could be curbed if such measures were introduced. However, new problems would arise if the cattle were allowed to reach full age. The current cattle population of 330 million would increase several-fold in just a few years. Damage to nature, animal epidemics and many other problems would arise.

What is stopping India from reducing its cattle population? The simple answer would be for people to stop drinking milk, thereby reducing the demand for it. Hindus, who have a vegetarian diet, consume far more milk and dairy products than other Indians. The increased milk consumption is one of the main reasons for health complaints. People in India who consume only plant food lead a healthier life than the meat-eating percentage of the population.

Milk production in India is based on traditional ways. The cows are usually separated from their calves in the evening so that the calves don't suckle at night. Early in the morning, the cows are milked and only then may the calf suckle the mother cow. This traditional method is cruel because the calves are not fed anything extra. The calves scream the whole night with hunger and thirst and try to find their mother. For this reason alone, the cattle in India are small and thin. This unethical method of farming should be reformed.

A reduction in milk consumption in India could protect the environment, prevent religious conflicts and finally promote a healthier life-style. The image of the sacred cow would remain for ever because it once helped the Indian peasantry. People could go on breeding cattle, but not under the conditions described above. Instead a cow could be kept as a pet.

If the Muslims of India would refrain from eating beef, they would be healthier and wealthier because they would not suffer from the dreadful diseases caused by beef consumption and would therefore not have to spend a lot of money on their health. If Hindus would accept that the milk from a cow should only go to feed its calf, the cattle in India would be much healthier and stronger. At the same time, people would be healthier, which in turn would reduce their medical expenses. In addition to the health and economic benefits, Hindus and Muslims could overcome their differences once and for all and a peaceful co-existence might be possible. Other benefits would be seen in nature. The tiger would return to the forests without approaching people and a significant proportion of arable land could be converted into forests.

Meat and the world population

Meat was always scarce in antiquity. Two kinds of meat were available to humans: first, game meat and secondly the meat of domestic animals. These two kinds of meat were scarce because the wild animals were mostly under the protection of authorities or noblemen and work animals were very rarely slaughtered. How difficult it was to hunt wild animals can be imagined from the medieval ballad 'Robin Hood'. Hunting was a privilege of the nobility and anyone else caught hunting was sentenced to death. Among domestic animals, it was only the older work animals or sick animals that were slaughtered. Due to the lack of fodder, animal diseases and the dangers from predators, animals were rarely bred for meat.

Until the beginning of the modern industrial age, the size of the world population remained constant, as was the case with other mammals in nature. Despite a sparse population, almost all parts of the world, with the exception of the Antarctic, were inhabited. Automation, which began in the first quarter of the

19th Century, is considered to be the start of the modern industrial age. In 1830 the population of the world stood at one billion. By the turn of the century, the world population had grown to 1.5 billion and up to this time, meat consumption was limited primarily to domestic livestock. The rapid deployment of modern firearms allowed for a worldwide increase in the availability of game meat, but after a short time this decreased. During the Second World War and for years afterwards, there was hardly any meat available from wildlife nor from domestic livestock.

By 1950, the world population had increased to almost 2.5 billion people and meat production to 45 million tonnes. Analytically it was 18 kg of meat per capita of the world population, but at that time it was only in the U.S, Canada, Australia and New Zealand that meat was regularly available to ordinary people. Those four countries had a total population of 175 million and it is estimated that they consumed approximately 40% of the meat stock.

The year 1950 is a turning point in population growth as well as the growth of meat production. In a 30-year-cycle from 1950 to 2010, the relationship between world population and meat production can be seen below:

Year	World population in millions	Meat production in millions of tons
1950	2.500	45
1980	4.500	125
2010	7.000	315

(Source: Chronology of UN Demographic Yearbook, New York and FAO Statistical Database, Rome)

Before the turn of the millennium, industrialized nations had reached the peak of an average of 100 kg of meat per capita. In countries like the US, the figure is around 130 kg meat per capita per year and the tendency is rising. The current 315 million tonnes of meat produced per year could mean an average per capita of 45 kg of meat for the world population. All countries in the world, except India, will gradually reach the 100 kg mark in the coming decades. Some countries such as the People's Republic of China will greatly exceed this 100 kg of meat per head.

Based on previous estimates, in 2040 the world population will have grown to 10 billion. With 100 kg of meat per head some 1000 million tonnes of meat will be required. If India were omitted from these calculations and the per capita consumption of China were increased, there would be still a need for more than one billion tonnes of meat. For a billion tonnes of meat, an average of 10 billion tonnes of animal feed will be required. In addition there is the demand for milk and eggs. Limited arable land, scarcity of feed, contaminated groundwater and more than three times the current amount of slurry will pose major problems. Three decades are not an eternity. This raises the question whether humanity with their meat consumption is preparing for a disaster or is there a way out?

Transition to a meatless diet

The daily consumption of meat begins early in the morning. Different types of processed meat products cover the breakfast table. A variety of boiled, grilled, smoked, dried, canned, heated or raw meat from different parts of the body of various animal species make up the first meal of the day. Lunch is often based on meat cuts such as steak, chops, tenderloins, or meat stew, sausages, meatballs or meat from a particular animal such as chicken, pig, turkey, cow or lamb. Dinner could be similar to breakfast or be a hot meal. If dinner is a cold meal, the meat products from the breakfast table are served again and in the case of a hot supper, the same meat range as was served for lunch re-appears. Accompaniments such as bread, potatoes, rice, pasta, salads and vegetables are regarded as secondary. Besides these fixed meals, there are snacks between meals: tea, coffee or other drinks may be consumed along with meat sandwiches or rolls, or a savoury sausage roll. Additionally, other animal products such as milk, eggs, yogurt, cheese, butter, cream, ice cream, etc. accompany the daily diet.

Modern man lives in total dependence on animals which are kept in barns and fed continuously. Humanity is very old but this way of eating is very new. However, people behave as if they have always eaten meat and they are firmly convinced that this is their basic right. Man, through his skill, is able to provide a wide range of meat cuts from the relatively small number of species which are bred for meat production, so it is unnecessary to look for other species to supply meat. A pig alone, from its snout to its tail, can provide so many different kinds of meat and meat products that man's appetite for meat is fully satisfied. For many consumers, a bratwurst is a quite different product from a steak, which in turn is totally different from liver sausage. This variety allows for different niches in the field of nutrition and consumers are convinced that they have a large variety to choose from.

For many people, life without meat is unthinkable. The first question they will ask is what they can eat if they do not eat meat. They regard a meat-free diet as poor and they worry that a meat-free diet will not provide them with the necessary amount of protein and vitamins. They worry about the economy: what would happen to farms if the meat industry were to collapse? They think of the variety of food with meat, which would no longer be available. These thoughts find no end, and finally they choose the easy option "after me the deluge" and continue to consume meat.

Deliberately or not, the growing world population has no alternative but to stop eating meat. Meat is no longer a privilege of the rich of this world, but rich or poor, developed or underdeveloped, people all over the world are gradually beginning to regard meat as a staple food. Sooner or later, the forests of our green planet will be cleared completely and the food crops will be replaced with fodder production. In some countries, such as Denmark, it has already come to this. This small country, with a total area of 43,000 square kilometers, has the largest per capita meat and grain production in the world. Denmark produces about 1700 kg of grain and 430 kg of meat per capita per year. As a result, Denmark possesses scarcely any natural forests, there are no more predators and the groundwater is largely contaminated with slurry and pesticides (CIA, World Factbook/Denmark, Natural hazards, Washington, DC, 2013). An increase in the number of pandemics and repeated droughts have been predicted and the consequences in the imbalance in the animal and plant world are unknown.

The origin of the words 'vegetable' and 'vegetarian'

The word vegetable in English

The term vegetable has no etymological roots and no meaningful prefix or suffix. However, when hyphenated, the word becomes: vege-table. The word "vege" does not exist in the English language. The second word table refers to a flat surface on which one can eat or write, for example. In various Indian languages the term for eating a meal is Voge, Bhog, Bhojan, or Vogen. In most North Indian languages, eating is Bhojen or Vojen (भोजन) in Hindi, and Bhoj or Voge in Bengali (ভোজ). In traditional Indian cultures, a table is not used for eating, but rather a Mej or a particular hallway or floor. For this reason, the English word table has remained in Indian languages. In the 18th Century, when the British colonialists began to introduce English customs into India in their daily lives, their employees took over English terms that did not exist in Indian languages. For example, the British dining table has been described as Voge-table, or Vej-table in India. In this way the Englishman living in India changed the name for the dining table into Vegetable. The table in the workroom or office was still called table and the table in the dining room became vege-table or dining table. Because the food on an Indian dining table was largely of plant origin, a new term for plant food became vegetable or dining table with plant food. This expression came to light only after the colonization of India. William Shakespeare and other English writers did not use the word vegetable. It surfaced at the end of the 18th Century.

European languages do not have a suitable term for plant food. The word Gemüse in German means edible herbaceous plant or parts thereof. The word légumes in French has the same meaning as Gemüse in German. In other major European languages such as Spanish, Italian and Portuguese the word for vegetables is derived from Verde, the color green, and so in these languages vegetables are called greens or Verdura. The collective term vegetable is very general, because each edible plant must be defined individually.

Vegetarian

The colonization of India by the British followed the takeover of the Muslim Mughal Empire. The food habits of Muslims and many other natives of India differ from those of Hindus, who prefer a diet of plant food. For administrative and tactical reasons, the British colonial powers were friendly with the Hindu upper classes and when they were invited to dine, they were served only plant food. The cooks were from the Hindu upper castes, for whom it is forbidden to handle and cook meat. To the Europeans, the Hindu elite of India were Aryans or Arian and the diet of Arians was a dining table of the Arian - called Vegetablearian or Vege-table-arian. The word table was shortened to T and so the word of Vege-T-arian means the culinary art of Arians. On 30th September 1847, when the first Vegetarian Society was founded in Ramsgate, England, no explanation of the name Vegetarian[3] was given. The term Vegeterian was translated directly into the German language. The word Arian is Arier in German and it has been added as a suffix, and so the food of the Aryans became Vege-T-arier or vegetarier. In other European languages the English word 'vegetarian' was absorbed into the national language with only a small change. So the English word 'vegetarian' is Vegetariano in Spanish, Portuguese and Italian, Végétarien in French and Vegetariér in Dutch.

It is widely assumed that the term 'vegetarian' means the dining table of the Aryans or the food of Aryans. This is not only misleading, it is wrong. Moreover, the word Aryan or Arian is a misnomer. Different racial theories have tried to prove that

---

3 There are many possibilities for using arian as suffix, but the prefixes of those terms have explicable meaning.

thousands of years ago, people, such as the Aryans from Central Asia, spread to the Indian subcontinent. From the end of the 18th Century to the first half of the 20th Century many of these theories, hypotheses and pseudo-scientific assumptions were presented in order to claim a better breed status. However, today's people from the Indian subcontinent do not call themselves Aryans. They define themselves according to where they live. So there are Punjabis, Beluschis, Tamils or Bengalis, but no ethnic group identifies itself as Aryans. Human beings do not need to classify themselves as either vegetarian, carnivore or omnivore, according to their food habit.

Regardless of religious influences, 'vegetarian' is a new type of diet. In the mid-19th Century, ideas about vegetarian diets came from Britain's former colony in India. Influenced by the food habits of the Jains and higher caste Hindus, many Englishmen took up a meatless diet for health or ethical reasons. The peaceful lifestyle of the famous politician Mahatma Gandhi impressed the Western World and his meatless diet also inspired many people. For health, environmental and especially for ethical reasons, many people do not eat meat but nourish themselves instead on plant-based food.

Veganism

Veganism is different from vegetarianism in that it eliminates all food products of animal origin such as fish, meat, milk and eggs from the diet. Donald Watson, the founder of the British Vegan Society, invented the name Vegan. Much like the British Vegetarian Society, which had been established in 1847 and given the name vegetarian without any explanation, so Watson introduced the name vegan and founded the Vegan Society in 1944. Watson explained that in consultation with his wife, he had selected the first three and last two letters of the word 'vegetarian' and come up with the term Vegan.

Donald Watson was certainly aware of the controversy over the term 'vegetarian'. He chose the prefix 'veg', meaning food or meal and the suffix 'an' meaning to belong, like American or Indian and put them together. Thus he removed the controversial concept of Arian or Aryans and used the new term Vegan instead, to mean eating a meal. Despite the different pronunciation, the expression Vogen in Hindi and the English term Vegan are closely linked. Watson removed the term Aryan or Arian from his definition because of what was happening in Europe during World War II, when Hitler's Arianism had mainland Europe in its grip. Watson thus tried to dissociate himself from the crimes committed in the name of Arianism. He also revealed in his famous interview with George D. Rodger on 15th December 2002, that he was familiar with the British-Indian connection. In this interview he used Hindi terms like "Raj" meaning 'rule over India', which are known only by experts on India.

Both vegetarianism and veganism, with their concern for health, ethics and the environment, are regarded as the most modern social movement of the industrialized countries. It would however be advisable to change the term 'vegetarian'. A simple solution would be to adopt the term 'vegan' and to further define it as ovo-lacto-vegan and pure-vegan. The philosophy of veganism is spreading gradually as a re-export and reaching a new generation that wants to monitor what it eats and take care of its offspring.

Definitions of meat in a selection of faiths

Rules and codexes about meat: whether eating it was allowed or not, and if so, what kind of meat could be eaten, first emerged in the context of mankind's various faiths. A policy is always associated with a ban and an excess or violation of the guidelines means sin, which in turn results in harm in life or after death. A punishment after death was meaningless, because death means the end of the body, and how can someone be punished in the absence of the body? Life after death promises eternal life. Life on earth is regarded as being temporary and life in the hereafter is considered to be eternal.

Jainism: Jainism is a pre-Christian religion and it is one of the first institutional faiths in the world. The main concept of Jainism is non-violence in the context of humans and animals. Because of their non-violence principles, Jains are not allowed to kill animals or eat the meat of dead or slaughtered animals. The Jains belong to one of the smallest religious communities, but at the same time they are among the most affluent in the world. Despite following a meatless diet, Jains show no health deficiencies caused by an abstinence of meat.

Hinduism: In Hinduism, there is no fundamental dogma prohibiting the eating of meat. However, the non-violence principles of Hinduism tend to correlate with a meatless diet. Because eating meat involves the killing of other living creatures, it is regarded as an impure product. For these reasons, a significant percentage of Hindus eat only plant food. Also the slaughter and consumption of the cow is prohibited. This attitude has something to do with the sanctity of the cow, which is considered to be the most valuable animal for agriculture and for the supply of milk. Because of the great diversity within the religious communities with a caste system, the types of food eaten by each are quite different. Many Hindus eat fish, meat and eggs; other Hindus do not eat any products of animal origin apart from dairy products.

Buddhism: The monks, who live on alms, have no choice about what they eat and for this reason some Buddhists eat meat, but they do not kill animals themselves. Otherwise, Buddhists are no passionate meat eaters. Because of the suffering involved, killing is strictly prohibited in Buddhism. Meat consumption is generally prohibited if the consumer watches the slaughtering, or if he orders it or if the animal is slaughtered for his own consumption. Also killing living beings for food purposes is banned.

The Abrahamic faiths: The three monotheistic secular faiths of Judaism, Christianity and Islam all have Abraham as their common ancestor. These homogeneous religious communities have a common creation story and they have the same geographic and demographic origin. Their basic livelihood goes back to nomadism. The arid to sub-arid geographical areas were difficult for seasonal agriculture, so many people wandered with their herds, which foraged for food. In a land where plants are not plentiful, meat is an important food source, provided it is available. In this case, the Abrahamic faiths recognized meat as legitimate food. However, the food codexes of these communities approved only ruminants with cloven hooves. Very strict rules are to be found in Judaism and Islam, in which the consumption of blood is strictly forbidden. However, it is extremely difficult to separate the blood mass from the meat. In Islam, an animal sacrifice is required once a year by people who can afford it. The meat of the sacrificed animal should be distributed mainly to the poor. Apart from in this situation, no one is required to eat meat. The strict regulations surrounding which animals can be eaten prevents excessive consumption of meat. At the same time the consumption of plant foods such as dates, grapes, olives and pomegranate is recommended.

Since the beginning of the 20th Century, most members of the Abrahamic faiths own a permanent residence. They practice various professions and very few of them still wander around with herds of animals. Despite the availability of plant food, they eat meat and claim it is their fundamental right to do so. As a result they are the ones who suffer the most from modern nutritional diseases. Their medical treatment seems to be more important to them than the struggle for daily bread. Christians and Muslims together form the majority of the world's population. They cause the most wars and they they fight with each other even within their own faith communities. They conduct most of the world's affairs, they initiate peace negotiations and they consume most of the world's meat.

If the members of the Abrahamic faiths reduced their excessive consumption of meat and other animal food products, they could certainly make a new and better start in life. The decline in livestock would mean the return of nature in which the safety of forests and wildlife is guaranteed. People would live healthier and more peaceful lives and their descendants would without doubt inherit a better world.

The different reasons for avoiding the consumption of meat

Many people avoid eating meat for a variety of reasons: religious, cultural, ethical and also out of conviction. The oldest reasons for giving up meat are due to health issues and out of religious considerations. Through education and awareness, more and more people worldwide are finding alternatives to meat in their diets.

Health care: Consumers these days have access to many different sources of information. Unambiguous proof of the harm caused by eating meat have led consumers to the realization that a meatless diet is healthier. The most important thing in life is health, and for this reason alone, many people from different classes of society have given up the consumption of meat.

Ethics: Civilization is based on ethics. The ethical doctrine in favour of not eating meat says: If other foods are plentiful, there is no reason to kill an animal or to eat a dead animal. Enslavement causes pain and mistreatment of animals also has no ethical basis. In many respects, ethical values are individual but they are also considered to be a collective responsibilty if it is a faith that requires a life without meat.

Environment: Worries about the environment and the future of the ecosystem are the primary concerns of the active world population. Land use, deforestation, fresh water contamination, depletion, harmful agricultural chemicals and the decrease in wildlife populations are directly connected to modern meat production. For these ecological reasons alone, conscientious people avoid eating meat.

Ailments: People who like to eat meat, often need to adjust their consumption of meat for health reasons. People who suffer temporarily from infectious diseases, such as influenza or gastrointestinal ailments, begin to eat meat again after their recovery. In contrast, poeple suffering from long-term chronic diseases such as diabetes, gout or heart disease often have to give up eating meat in the long term. Patients suffering from certain diseases, can be fed soup made with meat or bones and in the colder northern countries, the Inuit or Icelanders use the meat of the Greenland shark or seal meat as remedies. So far, no healing power in any type of meat has been proven and the modern northern people nowadays prefer antibiotics to the meat of rare animal species. Moreover, meat is not a remedy for a sick person because it is difficult to digest.

Dental diseases: Modern dentistry is a very new development and it is well-established only in the industrialized countries. Dental treatment is expensive and more than the half of the world's population is poor. Dental treatment in developing countries is mostly available to the substantial urban population. It is estimated that due to lack of dental hygiene and preventive treatments, more people worldwide suffer from dental problems than any other disease. Because of toothache or missing teeth, many people do not eat meat. In many industrialized countries, where dentistry is largely available, many people do not go for dental treatment. This is mainly due to a dental phobia and additional costs which the patients must cover themselves. Owing to their poor dental condition, many people, even in industrialized countries, where meat is very cheap to buy, cannot chew on a piece of meat. Instead, they eat minced meat. Industrial minced meat is a major chapter in itself and people in almost every state of health are able to swallow this mass of soft flesh.

Meat allergies: Meat allergies occur in children and in adults. They may be caused by different types of meat such as lamb, beef or chicken. Furthermore, certain proteins cause acute symptoms, anything from wheezing to anaphylactic shock - a disorder of the immune system. The appearance of meat allergies is particularly complicated if the person is outdoors. As a precaution many people suffering from meat allergies do not eat meat outside the home.

Disgust: Disgust is not only a socially-acquired cultural reaction, but also a personal physical aversion. Many people cannot imagine eating the body parts of an animal with relish. Many people cannot go to bed with body parts of dead animals in their stomachs, nor can they imagine a living animal as food. The thought of raw meat, blood, the act of slaughtering, the agony suffered by slaughtered animals or dead animals is scary and disgusting especially if the body parts of dead animals are chewed and swallowed. For these reasons many people choose not to eat meat.

Tastelessness: Meat, which consists of water, proteins and fat has no particular flavor. Without flavor enhancers, raw or cooked meat tastes very neutral. Therefore, many consumers find no reason to eat meat.

Smell intolerance: Meat smells of urine, feces, fodder, decay or has a particular odor. Many ingredients have been found to combat the natural smell of meat. Saffron, a very expensive spice, came to be used to cover the stench of mutton. To reduce a sexual fragrance, animals such as sheep, goats, pigs and cattle are castrated, a very painful procedure. Despite all these efforts to mitigate the smell of meat, the odor of each species remains. Through their sensitive sense of smell, many people develop an antipathy to the smell of meat so that they do not eat it.

Intolerance to meat: A significant proportion of the population can be classified as intolerant to meat. Since childhood they have shown an antipathy to meat. They refuse to eat meat without any external influences and if they are forced to do so, they may suffer from symptoms such as nausea, a rash or abdominal pain. The complaints may be caused by particular types of meat or meat in general. The symptoms are similar to a meat allergy but normally they are temporary.

Poverty: Meat is expensive because 10 kg of plant food are needed to produce one kilogram of meat. Only people in rich countries and rich societies in the world can afford meat regularly. If people suffering from hunger and malnutrition had a choice between meat and bread, the majority of them would certainly opt for meat. In developing countries meat is several times more expensive than plant food. Due to financial considerations, many people on low incomes forgo the consumption of meat.

Miserliness: Meat is generally expensive, except in industrialized countries. Because of its high price in many countries, miserly people choose to have a meatless diet. Food is a daily requirement and eating meat every day is an expensive habit.

Criteria for the selection of meat: People give up eating meat temporarily or for ever for a number of other reasons. The most important reasons are the meat that is on offer, slaughtering regulations and the hygiene measures. Pork or no pork, beef or no beef are decisions that are made about which animal can be eaten. Slaughtering rules in Islam and Judaism are of particular importance and a violation of these rules will lead automatically to the avoidance of eating meat produced in this way. Hygiene is an important factor in the meat industry. Lack of hygiene can cause a decline in the consumption of meat.

Provisional abstinence from eating meat: To reduce body weight, combat insomnia or due to a sudden change in diet, people sometimes avoid eating meat temporarily. The length of time varies; it may last a few days, months, years or it can even last a lifetime.

Shortage of meat: A decline in the supply of meat, for whatever reason, may also lead to an avoidance of eating meat. Shortage of animal feed, scarcity of grazing land and lack of financial resources for agriculture can lead to a shortage in the supply of meat. Animal epidemics, resulting in a larger number of animals being destroyed within a certain radius, can also cause a meat shortage. When certain animal diseases break out, the export of meat from that area is banned temporarily. This type of provisional abstinence from eating meat is not a voluntary act but a situation enforced from outside.

Animal cemetery: Without having a special reason, many people point out that they do not want to convert their bellies into an animal cemetery.

Are all non-meat eaters peaceful people?

Many people avoid meat for a variety of different reasons. They come from all classes of society. Peaceful as well as violent people can have a meatless diet. It is a personal choice and personal attitude, regardless of religious, political or social ideas. The backgrounds of people who do not eat meat differ, as do the backgrounds of people who eat meat. Whether a ruthless dictator who does not eat meat or a peace-loving person who lives on meat - there is no connection with these people's backgrounds. If felons decided to follow a healthy diet, many of them would reject meat as ordinary food. The argument "criminals are also vegetarian" is meaningless because criminals are also religious or family people.

Important incidents which can halt the consumption of meat

Abstinence of meat during a trip: Traveling or being on the road is often a risky undertaking. Eating, drinking, sleeping and answering nature´s call must be strictly observed. However, the most important activity is food intake. The food on offer during a trip is often of foreign origin and the modality of this food is often unknown. Unknown food can unbalance the body's health, which in turn can cause a premature end to the trip. Headache, abdominal pain, diarrhea, vomiting, dizziness, fever and sudden colds are the frequently-occurring travel diseases. Food poisoning, and viral and bacterial infections are the leading causes of motion sickness. According to the U.S. Center for Disease Control and Prevention (CDC) over 80% of sudden food poisoning are from food of animal origin. Viruses, bacteria or microbes are spread primarily through fish, meat and dairy products and therefore it is advisable to abstain from foods of animal origin during a trip.

Norovirus – the nightmare on a cruise: Crossing the ocean on a luxury liner, watching the sunset, eating lavish meals and having fun on a cruise is a dream holiday for many people. But the fear of becoming ill is often greater than the joy if the holiday-maker starts to think about a norovirus. A norovirus infection is a typical disease which can break out on a cruise. Thousands of passengers and many attendants are confined on board for weeks in the tightest spaces where food and drink play the most important role. In the middle of the ocean, noroviruses have no natural enemies and can spread relentlessly to all the passengers on the ship. Nausea, vomiting, diarrhea, stomach cramps, fever and headaches are the usual symptoms of a norovirus infection and can transform a dream holiday into a nightmare. The large buffet on board, the frequently-used toilet facilities, the on-board swimming pool and fitness centre, the railings or the friendly handshakes are the usual ways of contracting noroviruses.

Cruise companies underplay this viral infection, but nonetheless they take many hygiene precautions on board. Caustic chemicals, chlorine and disinfectants of all kinds are used to control the virus. Many sick travelers keep quiet about their unpleasant health experiences and so the plague remains in obscurity. According to empirical assertions, the prime suspect of an outbreak of a norovirus on board a cruise ship is the abundant variety of food from animal origin such as fish, meat and dairy products. In contrast, experience has shown that travelers who eat plant-based food suffer relatively less from this kind of disease. A cruise on which meat is not served seems impossible to imagine and it is as impossible to escape from the norovirus which travels on board as stowaway.

Meat in poorer countries: Modern meat production is a conversion process of plant food to animal food products. A large amount of plant food is needed to produce meat for human consumption. In industrialized countries, feed consisting of ordinary foods such as soy, canola, corn, wheat or fodder beet is the raw material from which meat is produced.

In poorer countries there is a lack of food for humans. If people do not have enough food for themselves, how can they feed animals with agricultural products? The common fodder for farm animals in the poorer countries consists of grass from the pasture and from agricultural waste. Pasture is barely present and so the animals remain thin and emaciated to the bone. These animals can hardly supply meat and because of old age the meat is often very tough. Nevertheless, meat markets exist in the poorer countries and the feeding of those meat animals is often a secret. Dubious waste can be processed into animal feed which cannot be controlled by anyone. A number of these dubious feeds are in circulation. Even the AFRIS, an FAO institute for animal feed research, recommends many deviant sources of feed such as wild animals, industrial waste, animal waste, animal carcasses, chemicals, waste paper, waste from the leather industry, etc. as suitable animal feed. For health reasons, short-term travelers to poorer countries should avoid eating meat.

Abstinence from eating meat on railways and on road transport: The number one cause of death in traffic accidents is fatigue. After a meal, the body becomes tired and the fatigue is more intense if the food contained meat. The microsleep is the same in humans as in carnivorous species, which suddenly close their eyes to sleep and wake up again shortly afterwards. Bus drivers, truck drivers and engine drivers have a special responsibility in traffic. The bus driver is responsible for the safety of his passengers and the truck driver for the safety of other road users. An accident with a massive truck is more devastating than with a small car and a tired engine driver can easily overlook a signal which can lead to a disaster. It would be safer if bus drivers, truck drivers and train engine drivers did not eat meat before and during their working hours. Also, the motorway service stations should not offer meat in their meals.

Meat, fish and aviation: Aviation is the fastest means of transport and it allows us to cross continents in a period of time during in which viruses and bacteria can easily spread. The airlines carry large amounts of fish and meat as cargo. Fresh products have a unique taste, a distinctive aroma and bring in a very high price which in turn can pay for the air freight. Freshly slaughtered meat and freshly caught fish are transported worldwide by air cargo and at the same time the whole range of viruses, bacteria and parasites fly along. On arrival the viruses, bacteria and other parasites may easily multiply. These kinds of fish and meat shipments are also responsible for rampant, worldwide zoonosis. These risky transports are carried out purely for economic reasons. The damage to humans and nature is regarded as insignificant.

Meat in catering on board an aircraft: Many people do not eat meat for religious reasons, out of conviction or for a number of other reasons. The number of meatless communities is on the increase. Many people can no longer stand the smell of meat, especially the smell of cooked meat. When hot meals on a long-haul flight are served, the smell of meat permeates the pressurized cabin. Some passengers keep their mouths and nose tight shut and some feel nauseous. The reaction is identical to dog meat being cooked in a closed room during which a dog-lover is forced to be present. Meat is a risky food which can lead to acute infectious diseases. For this reason alone, airlines should omit fish and meat from their menu in the same way as they prohibited smoking on board. Plant food on board allows for food variety, cleanliness, it is inexpensive and the different dietary restrictions such as kosher, halal, vegan or ovo-lacto would no longer pose a problem.

Halal and kosher in the air: The traditional three-class society is ubiquitous in aviation; there is First Class, Second Class and Third Class. However, the modified version of modern aviation describes these classes as the First Class, Business Class and Economy Class. From a linguistic point of view, this classification is incorrect because first is followed by second and third and not by the terms business and economy. On some flights there are two classes and more rarely, just one class. The so-called economy or third-class passengers make up more than 80% of all air passengers. As the name "Economy" says it is cheaper to travel in this class. The profit from economy class is extremely low and many of the airlines carry such passengers at a loss and therefore economy class is always associated with loss or additional costs.

On long-haul flights, special refreshments for members of certain religious communities are also offered. Kosher meat for Jewish passengers and halal meat for Muslims are available even in the economy class of many airlines. The Jewish kosher slaughter, called Shechita, is very complicated and only experienced people are allowed to carry out this job. After the shechita, a number of tests need to be performed to acertain whether the slaughtered animal was healthy or not. According to Islamic law, the head of the slaughtered animal must be kept in the direction of the Kaaba and a short prayer is to be spoken. If the rules are not respected, the meat is not kosher or not halal.

Most airlines usually offer only one kind of meat in the air - chicken. Beef and pork are rarely served to avoid any religious conflicts. In addition, lamb or goats meat are full of bones and also uneconomical. Chicken is the cheapest and most popular meat in the world, which can be handled from beginning to end on the assembly line. Compared to this, slaughtering according to the kosher or halal method is very expensive. A formal certification does not necessarily mean that strict religious rules have been observed. For this reason, it is recommended that believers should avoid meat in the air and for the length of their journey choose plant-based meals.

Consumption of meat by flight personnel: The job of a pilot involves precision work. Reliability, accuracy, and functionality are the key factors in aviation. A plane crash is often more devastating than any other traffic accident. For this reason, a pilot on duty, whether a commercial or recreational pilot, may not drink alcohol while working. Alcohol affects the nerves which can lead to concentration deficiency and fatigue and this can lead to dangerous situations. Despite a strict ban on alcohol, in many airline accidents, pilots show the same symptoms as associated with alcohol consumption. Overtired pilots can miss the runway, fall asleep and the airplane continues on autopilot or the pilots suddenly awake up shortly before landing. The causes of fatigue are attributed to stress, overwork or to health problems.

What makes a pilot tired in the air if he drinks no alcohol? Of course it is the food he eats, food with meat. The effects of eating meat begin a few hours after ingestion, resulting in fatigue. After a meal, all carnivorous mammals become tired and lie down to sleep. A stomach full of food without meat may also lead to fatigue, but the way this occurs is different. Carnivorous animals have certain stomach enzymes that help with the digestion of meat. Human stomachs do not possess these enzymes. So in order to digest meat more energy is used. Therefore fatigue is more common in humans than in predators.

The cockpit is small, comfortable and warm. Flying in the emptiness of the sky is monotonous, no one from inside or outside can watch and the two pilots sit there for hours with poultry meat in their stomachs. According to an aviation safety rule, pilots on long-haul flights can sleep a maximum of 40 minutes and after waking up, they can take the next 15 minutes to rest before taking control of the aircraft. A 40-minute-sleep is certainly sufficient for carnivorous species because they fall asleep immediately. In the case of humans, it is dependent on the brain activity if a person falls asleep immediately after lying down. According to the Federal Aviation Administration falling asleep on the flight deck is a common result of fatigue (FAA; Sleep and Psychomotor). Sometimes both the pilots fall asleep and fly over the destination airport where they were supposed to land. A ban on eating meat by pilots on duty could certainly reduce the risk of fatigue. Professional pilots should consume no meat at least half a day before their working time starts. This is also a precaution against sudden infectious diseases such as gastrointestinal diseases, flu and colds.

Avoidance of eating meat before a special date: A special event is something out of the ordinary. This can be a performance, a peron's own wedding, a first meeting, participation in an event or a long-planned trip. Often one has to plan long-term for such a date and mostly these are no ordinary affairs. Despite careful preparation, there is always the risk of a sudden illness, so that this event has to be cancelled. Industry has long recognized this phenomenon and turned this into a major sources of revenue for insurance companies. To guard against such situations, for example, a travel insurance which covers cancellation, is taken out. The insurance covers the financial loss, but the disappointment following the cancellation of a planned dream holiday or a special event cannot be refunded.

In general, a sudden outbreak of a disease is caused by bacteria and viruses and usually involves an infectious disease. These pathogenic microbes often migrate through food of animal origin. The incubation period of such contagious diseases lasts from a few hours to a few days. However, the latency (healing time) of these diseases is much longer. Depending on the importance of the event and to avoid risk, one should not consume fish or meat at least two weeks before the date.

Even if someone is not afraid of sudden illness, he should still not eat meat before this date. Such abstinence provides many benefits such as reduced fatigue in conversation, the ability to keep a cool head, less risk of anger, less thirst and less mouth and body odor. Although these things and phenomena do not necessarily occur, the probability of negative side effects as a result of eating meat is greater.

The avoidance of meat due to some precautions

People who are important: Those who are important in the family and for society are very notable people. From a breadwinner of a small family up to people with a lot of responsibility all belong to this category. The importance of these people will be felt if, for any reason, these people cannot exercise their role. A wage-earning family man, an experienced surgeon, a politician or an industrialist are valuable people, who might also be very important to society. The reasons why these people may suddenly be unable to perform their jobs vary. Nevertheless, the most common causes lie in sudden or long-term diseases. The causes of most diseases can be traced to diet and the causes of over 80% of diseases are due to foods of animal origin. People who feel concerned and recognize their responsibilities should refrain, at least for the period that they are at work, from eating meat. This reduces the risk of disease and at the same time conveys a sense of security.

Professional life with less use of muscle power: Professions which are done mainly sitting down, may cause long-term health damage if no regular physical exercise is taken. These types of jobs are often better-paid, which allows for the regular purchase of plentiful food. This food is mainly of animal origin, which the higher earners can well afford. As the years go by, they begin to feel somewhat uncomfortable in their own bodies. Insomnia and many other discomforts are directly connected to diet. To combat this, sports activities such as jogging, cycling, swimming, weight-lifting, hiking, running marathons and many other sports are practiced, and this also reduces the formation of fat in the body. In some cases, these activities are carried out to excess but with little success. It is the meat in their bodies which makes life difficult and eventually people go to the doctor. The physician identifies the one or other symptoms of food disease. Sports, drugs and fasting remain as unpleasant options for the rest of their lives.

Even proponents of meat-eating suffer. The so-called meat farmers, butchers, meat traders, meat haulers and chefs who must prepare meat for consumption, are also directly affected by those symptoms. Sports and medical treatment may be helpful temporarily, but in the long term there is no other solution except abstinence from meat. Doing sport can result in weight-loss by reducing the amount of fat in the body, or the muscles become tighter, but often the symptoms remain. It is mainly meat and other animal products that make life difficult. A plant-based diet combined with some leisure activities would lead to a better life. The most precious thing we have is our health and if our health is affected only because of meat, monetary wealth is of little importance.

Life with limited resources: People who have little income but want to live a better life, should voluntarily forgo meat and the consumption of other animal products. In this way, they can at least lead a healthier life. Their nutritional and medical expenses would in general remain very low and they would have more money available for other things.

The most important basic necessities of life are food, clothing and shelter. The issue of housing is connected to the prevailing climate. It also means the colder the temperature, the higher the energy costs. However, this is a feature in the industrial nations and it is very different in the warmer and colder regions of the earth. People living in these regions spend money on clothes less regularly and the clothes are worn longer. But the daily expenses on food are stressful for people with little money and it is made more stressful if they have to spend money on buying meat regularly.

Some economic benefits of the renunciation of meat: The production of meat costs more than the production of plant food because meat requires feed, fresh water, time, care and other related expenses. Additional expenses may be incurred for the control of animal epidemics, such as the mass killings and disposal of breeding animals if this becomes necessary as a precaution. All these liabilities make meat very expensive, but the agricultural subsidies of developed countries hold the price of meat down. These government subsidies come from tax revenues which the consumers must pay themselves. Intentionally or not, the consumer is forced to finance the production of meat. Nevertheless, giving up the consumption of meat will have many financial benefits, especially when entertaining guests, family and friends with plant food rather than meat. This allows for a greater diversity of food, ensures hygiene, it is inexpensive and it is a good deed for mankind and nature.

Meat consumption increases the use of unhealthy ingredients such as salt, fat and flavor enhancers. In addition, more cleansing agents, detergents and disinfectants are required in households where meat is eaten regularly. Meat consumption increases the body temperature and after digestion reduces the body temperature. These body temperature fluctuations increase heating costs in the winter and cause discomfort in summer. A winter spent without meat is healthier and cheaper because a meat-free diet reduces the probability of catching winter colds.

The consumption of meat also produces a larger amount of waste, be it finished products or unprocessed meat.

Meat and spiderwebs: The natural decomposition in an animal is activated automatically as soon as it dies or is killed. The bacterial degradation results in odors from decomposition, which attract other animals such as flies, insects or other scavengers. Flies come not only to eat carcasses, but also to lay their eggs, which in turn develop into maggots which help to decompose the flesh of the carcass. Great health and safety measures are needed to keep these animals away. Nevertheless, scavengers of any size have a desire for meat. In a household where meat or fish are eaten, various species such as flies and insects are found. Spiders come into a house to catch these animals and make their webs. Cobwebs trap not only flying insects, but also dust and fuzz and make an ugly picture in living rooms. In some parts of the world, spiders can also be poisonous. Spiders have little to do with humans except through the connection with carcasses. Flies, which are attracted by ripe fruit can also be the target of spiders, but the effect is negligible. Spiders are always present in the household, but more so in households where fish or meat are eaten.

Meat and beauty care: Carnivorous mammals often have pretty fur. The skin beneath this beautiful fur is not smooth, but spongy and ugly. In addition, carnivores have a strong body odor caused mostly by eating cadavers. The odors of the respective animal species are different. The smell of a billy goat is much stronger than that of a pig. The scent of musk deer is even marketed as a perfume. Even without sexual fragrances, predators have a much stronger smell than herbivores. Excessive meat consumers frequently use skin creams, perfumes and other body care products. Regardless of culture, the cosmetics market in those countries where meat is the most important food is very large. A reduction in the consumption of meat would also reduce unnecessary expenditure on cosmetics.

Meat in canteens: Wealthy countries have wealthy universities where at lunch a variety of meat is served. A stomach full of meat results in afternoon fatigue; weary professors make their way towards yawning students who doze off during lectures.

Fatigue is also the cause of many industrial accidents. In the canteens of many large industries mainly meat-based foods are served. The meat content of lunch makes workers tired which is why most work accidents occur in the early afternoon hours. The omission of meat in the diet of people doing risky jobs is a better precaution against accidents than the use of preventive agents against fatigue.

Eating no meat does not mean that one cannot get tired after eating food of plant origin. Every meal resulting in a full stomach causes fatigue, but the digestion of meat and vegetable food differ from each other. This phenomenon can also be observed between carnivores and herbivores in nature. Carnivores frequently take a nap after eating. In contrast, herbivores eat their food throughout the day. They sleep for a long time, but not as frequently as the carnivores. A short sleep-like fatigue is the cause of many accidents.

Meat of unknown origin

Puzzling over the origin of a piece of meat has no limits. Even if a piece of meat can be traced to a particular animal species, still the modality of this meat can remain hidden. Meat of unknown origin is handled not only by dubious meat traders, its presence is ubiquitous. Criminal activities are to be found in all sectors of the economy and the meat industry is no exception. Humanity has been engaged for a very long time in this complex issue of meat of unknown origin. An example is the Jewish dietary code which prescribes a set of rules in order to render the meat as kosher. First, a suitable animal has to be selected, then it has to be slaughtered according to kosher rules and then rules have to be followed for disease scrutiny whether the slaughtered animal was healthy or not. Finally it may be ready for preparing dishes and all of this is possible only if the origin of the animal is known.

Meat scandals accompany the history of mankind. The secret acts of the meat industry lead to scandals which range from small, individual meat scandals to problems at national level. A method for determining the origin of meat is not available. Meat ingredients in convenience foods, meals prepared with minced meat, very different-tasting meat or meat that is suddenly sold very cheaply may also contain meat of unknown origin. Game meat may also be of concern because the diet of the hunted animal remains unknown. However, no reliable data is available to identify meat of unknown origin.

Russia stopped the import of meat from Germany because it claimed that meat from third countries such as China and Pakistan were being imported to Germany and then exported to Russia as German products (RIA Novosti, Moscow). This type of label falsification is a part of the modern world meat trade.

Consumers have confidence in reputable meat manufacturers; they are convinced that the meat comes from a reliable source. According to an old saying "Trust is good, control is better". But in general it is impossible to control finished food products containing meat. Moreover, the modality of fresh meat from afar is also unknown and uncontrollable.

Professional sportsmen and the consumption of meat

Sports activities are body movements that are primarily practiced in leisure time. Sport can become a profession if the sportsman earns his livelihood from it. Therefore, this profession cannot be separated from other professions which involve physical exertion. Nevertheless, the activities of a professional sportsman are also referred to as sport. The crucial element in sports activities is the stamina of the body. Weakness, fatigue or tiredness impair sporting activities. Physical endurance was very important to the survival of primitive man when he lived in the wild. To save himself from enemies or in search of food he had to run long distances or he had to jump and climb quickly. In this case fatigue or exhaustion could mean death.

How can stamina be built up? The answer is through sport. This is difficult, because increasing your endurance through sports is a lengthy and tedious process. A better answer would be the diet. Instead of debating which diet is better than others, the diet of different mammals can be observed. The following is an example of physical endurance and diet in two different species of identical body size:

The phenomenon of the horse and lion: The horse is one of the most beautiful animals on our planet. It has great strength and stamina, it can climb mountains, cross rivers, gallop in the desert or snow and it can carry heavy loads. It is a very quiet animal, which sleeps predominantly in the standing position and it eats plant-base fodder. In contrast, a lion is very strong, yet very vulnerable. It has very little stamina and other than to hunt its prey, it takes part in no other physical activities. It is very aggressive, it does not share the common prey with the pack from the beginning, it sleeps over 18 hours a day and eats only meat.

The meat diet enables a lion to develop enormous strength with which to catch an animal within a very short-range. But a lion has no stamina and can only chase an animal for a short period of time and if the lion tried to do more, it would die of exhaustion. Natural man nourished himself mainly with plant food and rarely with meat. If he had eaten meat regularly, he would have often been tired and this would have constituted an extreme risk because human anatomy is suitable neither for defense nor for offensive purposes.

Above the Arctic Circle, only the polar bear can be described as dangerous to humans. Polar bears rarely attack humans and so the meat consumers in the north were hardly ever affected by this menace, whereas in other areas, cat-like human enemies were widespread.

The modern sportsman takes after a lion rather than a horse and as a result suffers from exhaustion, pain and takes risks with his life. A football player, a marathon runner or a weight-lifter eats like a lion, leopard or tiger rather than like a horse, bull or elephant. Sportsmen can certainly deliver a better performance and have more endurance if they eat plant food rather than meat. But the time is not ripe for people to accept this truth. Thousands of sudden cardiac deaths occur every year among sportspeople such as football or basketball players, runners or multi-discipline atheletes. Sports physicians trace the causes of sudden cardiac death to heart disease, heart rhythm disturbances and physical stress. But they are not convinced by the argument that meat is bad for sportspeople.

Meat in hospitals

Hospitals carry health risks for sick as well as healthy people. Those places are full of viruses, bacteria, fungi, spores and parasites. Patients with infectious diseases, with surgical wounds and injuries or people with poor health are very susceptible to catching other diseases. To combat these dangers, constant cleaning with disinfectants takes place. Despite all these precautions, the pathogenic micro organisms attack from all directions and cause a large number of casualties every year.

Infections picked up during a hospital stay, also called nosocomial infections, may originate from the use of medical equipment, in the operations performed, in the lack of hygiene and also in weaker older patients. Cleaning and constant washing are practiced as a prophylaxis. However, a decrease in hospital-acquired infections is not in sight, but more and more resistant bacteria and viruses have been found in the hospitals.

Similar to the scavengers lurking in injured, sick and dying animals in the wild, viruses and bacteria lurk in the environment of hospitals. The precautions taken in the wild to keep the scavengers away involves leaving no traces of blood or body parts and covering open wounds. Modern hospitals are perfectly safe when it comes to this kind of hygiene. They leave no open wounds or traces of blood. Yet they approve the distribution of large amounts of body parts of other animals on the hospital premises. These body parts, called meat, which include traces of blood from other species, are served in hospitals as daily food for the patients. The key causes of hospital infections are due not only to the concentration of sick people in one confined place, but also to the serving of meat. This allows the the bacteria and viruses to go to work. It does not matter if the meat comes from a cow, a pig or from a wound, the microbes will find their way there.

Many hospitals have their lunches delivered from a commercial kitchen. Certainly the microbes also use this delivery service to enter a hospital. In addition, hospitals have their own kitchens, where meat, sausage, ham, salami, eggs and dairy products are prepared. Products of animal origin are extremely susceptible to microbial contamination. The presence of animal food products in a hospital supports the possible contamination by viruses and bacteria.

More than 70% of hospital infections are caused by bacteria. Bacteria multiply rapidly in perishable foods such as fish, meat, eggs and dairy products. Due to growing prosperity, the consumption of meat in hospitals is increasing. In a hospital, where there is a greater risk of bacteria and viruses spreading, it is hazardous to supply the patients with meat. The labelling of meat with an expiration date is a biological contradiction. Meat, or the muscle tissue of an animal, does not decompose if the animal is alive and cold storage does not mean that meat remains intact. The decomposition bacteria are active at almost any temperature. Otherwise, above the Arctic Circle animal carcasses would never decompose.

Apart from hygiene management, what are the other options for avoiding hospital infections? The microbial world is new to us. We know very little about its diversity, potential magnitude, velocity, life cycle or its areas of retreat. They are stable, strong, powerful and merciless. Their method of multiplying is so complicated and they are so small that they will find protected zones even on freshly-cleaned surfaces or recently-washed hands.

A modern hospital uses a lot of cleaning agents. Corners, edges, arches, all unreachable locations are carefully cleaned and disinfected. A monotonous cleaning procedure in hospitals is the disinfecting of hands. Hands are the most important organs of the body and these are exposed to constant rubbing with disinfectants in a hospital. Before any treatment involving the use of hands, the hospital staff must rub their hands with chemicals that kill the bacteria, viruses, spores and fungi. It is not known to what extent human hands are tough enough to withstand this kind of chemical attacks without a negative effect.

More and more unknown microbes are emerging and humanity must continuously take antibiotics, inject vaccines and suffer through the use of disinfectants. A voluntary renunciation of meat and other foods of animal origin in hospitals would be a better way to combat infectious diseases than other offensive or defensive measures.

Voluntary renunciation of meat for health professionals

Doctors know the various ways in which a person can get sick. They know that it is not easy to fight globally-present diseases with the profit-oriented remedies that are on the market. Doctors want to cure their patients and send them home healthy but it is a hard work trying to save lives when the chances of success are not guaranteed.

Needless to say, man can become sick because of his diet. The modern physicians know what kind foods can cause certain diseases. Food of animal origin are the most popular and most commonly eaten food of an affluent society and these are at the same time responsible for numerous nutritional diseases. A medieval story from the Far East is worth a mention.

A mother brought her son to a guru to rid him of his excessive sugar consumption. After listening to her complaints, the guru advised the woman to take her son home and come back a week later. The woman returned with her son a week later and the guru kept him to work on a cure. When the woman returned to pick up her healed son a few days later, she asked the guru why he had refused to treat her son the previous week. The guru replied that he had also eaten too much sugar and needed a week to break this habit.

If doctors set an example by giving up meat, a lot of their patients and other people would follow their example. Fewer patients would mean less income and less growth in the medical industry, so it becomes a question of ethics. Which is better - more sick people and more income for doctors and the medical industry, or fewer sick people and reduced income? According to medical ethics, healthy people bring more joy than the sick.

Meat and legal defense

In order to escape starvation a person may kill and eat a lamb, but to escape the same starvation a wolf or a leopard may not kill a lamb, because the law was conceived only to protect humans and their property and not to serve wild animals.

Man is a part of nature and a healthy natural environment is the most valuable possession to humanity. Natural laws and man-made laws agree on this point: the environment needs to be protected. The modern meat industry and the excessive consumption of meat have been proven to act against nature. They violate the laws of nature and put the future of mankind at risk. Based on the following points, the industrial production of meat should be stopped:

1. If enough other foods are available, there is no reason to kill an animal
2. Livestock destroys the population of wild animals
3. Livestock destroys forests
4. Livestock causes air pollution
5. Slurry from factory farms destroys fertile land
6. Livestock contaminates the fresh water supply
7. Mass enslavement of certain animal species is immoral
8. Mass culling of young animals is cruel
9. Mass spread of zoonoses is dangerous
10. Fodder for livestock reduces plant-based staple food for humans
11. Meat intensifies cultural conflicts and prevents world peace
12. The consumption of meat causes innumerable nutritional diseases

For the benefit of humanity and nature, a new debate about a diet without meat needs to be initiated and laws need to be introduced. It needs to be legally and scientifically proven that the modern habit of eating meat is unacceptable.

Chapter V: Food of the future

Giving up meat as a staple food

Because of the growing world population and the increasing scarcity of arable land, the practice of intensive land use for the cultivation of cereals and legumes is not viable. These types of crops are increasingly being used as feed to produce meat, milk and eggs. Meat is the prime product and the other two commodities are merely by-products. Even Hindus, who do not eat beef and are strictly against the slaughter of cattle, are passive beef producers. They keep cattle on a large scale, treat milk as a precious food and when the milk yield declines, the cow is sold and finally consumed as meat. All animals which are kept for economic purposes are meat suppliers. Some of these species provide a single by-product and when this by-product can no longer be acquired, the animal is processed into meat.

Whether it is desirable or not, the days of meat as an ordinary food product are over. No country, no nation and no tribe has the collective right to destroy the common environment and the environment of future generations, merely to satisfy their craving for meat. This fact does not deny that meat is a staple or significant food for mankind, and it would be unjust and contrary to nature to demand a complete ban on producing and eating meat. Depending on circumstances, meat could still be produced as food, but not as a profitable economic product. In emergency situations, such as during a disaster, meat, regardless of its origin, can be a life-saver. Some religions even recommend their adherents to consume forbidden meat if it means saving a life. The delicate question of cannibalism can have no general application; each individual must decide for him or herself whether or not to break this strict human taboo. People who form a symbiotic relationship with farm animals, may also decide for themselves whether or not to eat an animal that has been a companion. The members of some Asian religions offer their deceased to vultures, because they believe that after death the body should be available to scavengers as food. The carcasses of dead animals are a common food source in nature. It should be left to people to decide themselves whether they opt for the consumption of dead animals or not.

Free will can be exercised in deciding whether eating meat is a natural right, but free will used to decide how to market meat has never been a natural right. Meat marketing should be exposed as one of the worst forms of trade among all commercial undertakings. Since the beginning of civilization, the animal world has been at the service of humanity and humanity has now evolved so far that it no longer needs the service of animals. Instead of living in a cave, human beings now live in skyscrapers, and instead of spending all day searching for food, there is a supermarket or a restaurant at their disposal. But the lives of farm animals have not improved at all; in fact they have become much worse. Animals that could once live for more than 20 years, are lucky if they reach five years today and animals that lived alone or in pairs in a stable, now live crowded into confined spaces. Human beings have become more civilized towards each other but much more uncivilized towards nature and animals. Giving up meat as staple food is the first step to freeing the animal world.

Save existing wildlife, not dinosaurs!

Some branches of science deal with species that have been extinct for millions of years. Dinosaurs belong to this fantasy world, in which giant herbivores and carnivores dominated the green landscape. Society invests billions of dollars to promote imaginary research into them. Active research, genetic engineering, fiction or filmmakers concern themselves with dinosaurs, which will never return. According to reconstructions of dinosaurs, these ugly beasts have nothing in common with today's wildlife. The wildlife of the human age is wonderful. The great diversity of feline animals, or herbivores, all are impressive animals and they are threatened with extinction. With the current trend of natural destruction it will not take long before predators like the big cats are classified along with the dinosaurs.

We do not know which other species existed at the same time as the dinosaurs and we do not know what the consequences would be if present-day predators died out completely. Modern meat consumption is disastrous and they are putting nature at risk.

It is because of meat that people refer to predators as enemies and exterminate them systematically. The publication of a Red List of endangered species or individual actions such as "Save the tiger" or "Save the leopard" will not help in the long run, because all of these actions do not address the heart of the problem, namely a ban on hunting in the habitats of endangered species. However, such a ban is also doomed to failure because at the edge of the protected zone, grazing animals such as cattle and goats are bred, and when farm animals are attacked by predators, hostility becomes inevitable.

Precautions for wilderness management

As a remedial measure, nature must be allowed to revert to its former state. An abandoned river landscape returns automatically to its natural state. Plants will grow again and wildlife will find its way back - without any human assistance. Waters should no longer be diverted for economic reasons. Be it inland waters, coastal waters or offshore waters, they should not be put to use intensively for profit, but should be available to nature and for common usage. Water should not be polluted with industrial waste, blocked, diverted, used for intensive fish farming or used for the disposal of effluence. The forests need no species of plants that were first grown by humans nor do waters need specific species of fish produced in breeding tanks. A control or monitoring of forests and waters is unnecessary. It is not the job of a forester or hunter to decide how many boars may live in a wooded area; it should be up to the predators.

Scattered settlements are contrary to nature. The inhabitants of the settlements need to protect themselves from wild animals. They have to withdraw from areas where predators roam and should not try to force wild animals to retreat deep into the forest. All newly-claimed areas of nature, which are under intense industrialization, should be largely rehabilitated. Large forests represent a large living space and contain sufficient food resources for the animal species that live there. The greater the wilderness, the less the risk to man from dangerous animals, provided that a safe distance between civilization and nature exist. People may not be afraid of the dangers in the wild, but they should not expose themselves to the wilderness, as this would provoke the predators that live there. The bait which attracts the predators is the meat produced from farm animals. No matter how well-hidden chickens, pigs and cattle are, the smell of feces and urine in the air will alert the animals in the wild of their presence. Once larger predators like big cats and wolves smell the presence of ungulates such as cattle, sheep, goats and pigs, their instincts compel them to pursue them. Smaller predators such as martens, wild cats or raptors are attracted by poultry such as chickens and ducks. Thus animal breeding is a pre-programmed conflict between man and nature, but animal breeding without the wilderness means a threat to the whole of nature.

Now to come back to the question: Wouldn't a return of predators to the wild mean that they would attack people in remote villages and towns? This question cannot be answered with a decisive yes or no, as a number of conditions are associated with it. Where animal husbandry is practiced, an accumulation of human excreta is not noticeable. Anyway, predators prefer four-legged hoofed animals, which they hunt in the wild. Pets such as dogs and cats are of no interest to predators. Omnivores, such as bears, approach the suburbs of settlements, attracted by the smell of leftovers in waste containers, ponds full of fish or bee hives. Farm animals attract predators, but a dead animal in the form of meat attracts countless small or invisible species which can be a worse danger to man than wolves or tigers. However, people are organized and know how to protect themselves. Even if a predator attacked a human, man can arm himself but a predator cannot. Hazards are an unavoidable companion of life and taking precautions is the only way to stay alive.

Cutback in the cultivation of maize, wheat, rice and soy

Cereals such as maize, wheat and rice make up about 60% of the world's arable land. Cereals can be harvested every few months and, depending on climatic conditions, they can be grown from one to three times a year. Cereals are among the most intensively-farmed crops in the world. The land has to be plowed, fertilized and often watered before planting. As a result of this intensive agriculture, soil is often lost through erosion. Chemicals used on plants are carried by rain or by irrigation into the groundwater. In addition, cereals no longer serve their original purpose of providing man with a staple source of food. Apart from rice, other cereals such as maize and wheat are mainly used as animal feed. If food of animal origin were gradually reduced, there would be no reason to produce such large amounts of cereals at an enormous environmental cost.

With the decline in cereal acreage, a surplus of arable land would become available, which could be converted back into forest. As a result, at least 80% of the artificial watering used in cereal cultivation would no longer be needed. In addition, the contamination by chemicals used for cereal cultivation would disappear. The many other polluting factors that are associated with the cultivation of cereals would also be minimized. Cereals should go back to serving as a staple food for human, not as fodder for farm animals, for whom cereals have never been a natural food source. Climate-dependent reduced acreage could deliver better and healthier cereals than those resulting from overproduction. History teaches us that a total dependency on cereals can lead to hunger and malnutrition. With a view to the growing world population, humanity must wean himself off this addiction to cereals and learn to eat other products of the nature.

Apart from cultivating cereals, man is also increasingly focusing on the cultivation of leguminous crops. One of these leguminous plants is soy, which in the last three decades has been increasingly grown in huge quantities. Soybeans are grown primarily as animal fodder, which is used for the production of foods of animal origin. A decrease in animal husbandry, would reduce the demand for soybeans, which would benefit the jungles in South America in particular. If the world cereal production were reduced by half, the patchwork appearance of the earth would change and it would take on a more satisfying aspect.

Recognition of other staple foods

There are many other plants, apart from cereals, which could play an important role in human nutrition, depending on climate and soil conditions. Most of these staple foods require little irrigation and the cultivation is in many ways environment-friendlier than the cultivation of cereals.

Potatoes: The potato is a staple food grown around the world, which requires very little water, heat, nutrients or care. The potato grows on all continents, from as far north as Alaska to

New Zealand and it thrives in all regions and climatic zones of the world. The potato plant can withstand temperatures between 0° C and 50° C, it is not sensitive to strong wind, strong sunshine or heavy rain and the plant is not a neophyte which displaces other plants. Unfortunately, this precious tuber is cultivated only on one percent (18 million hectares) of the world's arable land.

Using traditional methods of cultivation, only one ton of grain can be harvested on one hectare of land, but using the same methods of cultivation, one hectare of land can yield about 12 tonnes of potatoes. The potato plant has no special soil requirements and can be grown in pots, in small gardens, in fields, on beaches, on roadsides, along railroad tracks, along runways, in the steppe, in semi-deserts as well as on high mountain slopes, always producing a successful harvest. Potatoes are easier to store than grain and in many ways the potato as a staple food is healthier than any cereal. If the potato, along with grain could find recognition as a staple food, as it is in Germany, Switzerland or in the Netherlands, the total demand for farmland would be reduced by at least half and at the same time there would be an end to food shortage around the world.

Yams: The yam is a vine plant that grows in all the tropical regions on earth. It can grow successfully in the wild, requiring almost no care and provides a normal harvest of 10 tonnes per hectare. There are several hundred different species of yam and a single yam can weigh from a few kilograms to over 70 kilograms. Yam serves as an excellent staple food in many warm and rainy regions of the world. Because the yam is very durable and can be stored easily, this root could play an important dietary role.

Manioc: Manioc or cassava is a tropical plant which can produce about 12 tonnes per hectare. The plant can grow to be two meters high and it can be grown as a monoculture or as a companion crop. The cassava plant needs sun, heat, humidity, but it needs no care, no irrigation, no fertilizers or pesticides.

Sweet potato: The sweet potato is a tropical plant which can produce up to 14 tonnes per hectare. Cultivation is easy, storage has no particular requirements and the crop is also edible in its raw state.

Plantain: The plantain, also known as the cooking banana, is a staple food in many warmer countries. The banana plant requires mainly rain, sun and heat and a successful harvest yields up to 10 tonnes per hectare.

Other major staple foods: Taro, papaya, pumpkin, dates, peanuts, beans and peas are significant staple products. Depending on the climatic conditions, these plants grow without any special human care, unlike excessive monocultural farming. Plants such as chayote, taro and sorghum grow wild in certain areas without any special care, and if these were introduced more widely into the wilderness, they would provide a safe food source.

Companion planting: All of the above-mentioned staples can be grown as companion crops. Companion planting, also known as intercropping, resembles a park and not the monotonous fields of cereal crops. Companion planting is perennial, the area is always green, it looks like a wilderness and the harvest is regular. In temperate climates, potatoes and the other regionally grown food crops could take on the same role. Grains could still be grown as a staple food in limited quantities where the plant grows successfully without irrigation and artificial chemicals.

Permanent crop: A permanent crop of any kind is better than short-term harvests involving the intensive plowing of land. Fruit trees, nut trees or the smaller taro plants are far more appropriate than intensive agriculture as animal feed. Long-term, permanent plantations of a monoculture are in many ways better than a three-month grain crop. It is high time that the world population fed itself from permanent crops and tuber crops such as potatoes, yams, taros rather than from foods of animal origin. At the same time, the establishment of a permanent crop, adapted to global climatic conditions, would be more appropriate than development aid or famine relief.

Companion planting of coconut palms, oil palms, date palms, bananas, papayas and pineapples, or the intercropping of apple, pear or peach trees is better than farming a monoculture. In intercropping, pests and plant diseases do not affect the entire plantation, but only a certain species of plants and thus preventing a total crop failure.

Sun, heat and fresh water

Sunshine, heat and fresh water are the most important preconditions for the growth of crops, but these are not always available together. Water can be regulated, but not sunshine. For this reason artificial irrigation is implemented in warm, sunny areas. Irrigation is a very risky procedure in hot, sunny areas because the water resources in such areas are scarce and very sensitive. Export-oriented cultivation using artificial irrigation has resulted in the drying up of many inland sources of water. The disaster of the Aral Sea is a good example of the results of irrigation in arid and sub-arid regions.

Many countries with arid to sub-arid climates cultivate export-oriented food crops using the existing groundwater or any available inland water, thereby exposing the ecosystem to risk. The cultivation of citrus fruits in the Mediterranean is wholly based on artificial irrigation, which in the long term could cause a natural disaster of unimaginable magnitude. Irrigation is only reasonable for the cultivation of locally-consumed plants and not for export to an ever-increasing population abroad.

Food production should be geared to rainfall. The world receives enough rain to produce a sufficient amount of food. It is the duty of mankind to grow climate-related plant species. It goes against nature, and could even be termed an eco-crime, if people who have enough rain and land to produce food for local consumption, feed themselves with imported food produced with the help of irrigation. Rain is the main source of fresh water supply, whether for drinking purposes, for irrigation or for storage.

A global mandatory labeling of foods with irrigation information should be introduced, for example a 'Rain-fed' label which must go hand in hand with strict monitoring. Precise details of the production areas would also provide information on annual rainfall in that area. Rain-fed agricultural food products should be rated better than organically grown foods with artificial irrigation. This drastic step would help to rehabilitate the fragile ecosystem of the desert and semi-desert areas.

Abuse of foodstuffs to produce conventional energy

Food provides energy for the body, which enables life. This energy is released as a result of the digestion process, which must be continuously in operation. Because the existing fossil resources will one day run out, man has had the idea of using these same energy sources, in the form of plants grown for human and animal consumption, to produce conventional energy for mobility, heat generation and industrial processes. Cereals and oilseeds are converted to gasoline, diesel or electricity without thinking about the consequences.

To produce these types of energy, large amounts of plant foods such as cereals and oilseeds are needed. But the world does not have enough arable land to grow sufficient food for its population. If half of the world's population suffers from hunger, how can plant food be utilized to as a source of conventional energy? Half of the grain harvest is fed to farm animals and now man has also begun to extract energy from this grain for use as fuel and in industrial processes. A large part of the waste resulting from energy production is fed to farm animals. This feed is contaminated with chemicals and enters into the human body through the food chain. Food scandals will never end because the demand for conventional energy and the demand for food of animal origin are on the increase. The energy obtained from plant food is always smaller in relation to the waste produced and the use of such waste as a recycled product is then used as a food source for humans. So, rather than consuming wheat as bread, it is first converted into ethanol to drive a car and then the waste is fed to farm animals in order to produce milk, eggs and meat. What is more, the fields on which cereals are grown are fertilized with the excrement of farm animals. An excellent utopian concept that can only end in a disaster. Eroded land, contaminated fresh water and polluted air, followed by hunger and disease will afflict the world.

There is only one justifiable reason for the cultivation of food, and that is to feed mankind. The other uses can be considered as abuse because they work against nature. The animal world does not need food produced by human if they are not enslaved and food plants do not exist so as to serve as a source of energy.

Foods which could replace meat

Many people who have become accustomed to eating meat and are now looking for a way out, can find a large selection of plant products which resemble meat. Numerous industrially-produced meat substitutes have the same the properties as meat in terms of consistency, structure, bonding, smell and taste. These meat substitutes allow for a meatless 'meat' diet. These products can help in the early stages of changing to a meatless diet. As time goes on, people get used to other foods and the need for meat substitutes becomes less crucial than in the initial stage.

From the chemical perspective, the common types of meat contain an average of 76% water, 23% fats and proteins, less than 1% carbohydrates and some minerals. From the human perspective, the main ingredients of meat are blood, skin, bone and connective tissues. Because of the high water content, meat tastes watery in its natural state. Odor and texture impart information about the origin of the meat. The watery taste of meat in its natural state and the smell of urine is not appetizing. For this reason meat is prepared in many different culinary ways.

If, instead of meat, meat substitutes of plant origin are used in a meat recipe, the dish will be identical in smell and taste as the original meat dish. The spices that are intended for meat dishes assume a significant role in the preparation of plant recipes. There are a number of plant foods that can be used as meat substitutes and prepared in a great variety of ways. Vegetables that can be used in meat recipes have a solid consistency, such as aubergine, banana-flower, plantain, chayote, green papaya, hokkaido, butternut pumkin, turnips or broccoli. There are many varieties of aubergine and in the last decades it has won more and more importance as a meat substitute.

Potato, taro, yams have a similar texture but can be prepared in different ways. There are many varieties of potato and the potato could replace meat in many ways. The hairy taro tubers have a firm consistency, are suitable for transport over long distances and do not require any special storing conditions for several months. Yam tubers consist of a very good pulp which can be prepared in various ways as a meat substitute. At normal room temperature yams can be transported and stored for several months.

A number of fungi and mushrooms possess an identical smell and taste to meat and many of them can be prepared as a meat substitute. Champignon, chanterelle, oyster, Sparassis crispa, beefsteak polypore or ox tongue, parasol, laetiporus, edible boletus, giant puffball, schiitake can be used in many different ways. The fungus schiitake for example has a meaty taste and the beefsteak polypore can be used as a substitute for steak.

Mycoprotein is a meat substitute derived from fungal cultures, which has an amazing similarity to meat. In huge fermentation-towers, oxygen, nitrogen, glucose, minerals and vitamins are constantly added to fungal cultures (Fusarium gramineurum). The Mycoprotein derived from this is prepared as meat substitute.

Some ferns can be used as a culinary substitute for meat. Ferns are flowerless plants. About 12,000 species of these over 400-million-year-old, primitive plants are widespread in many climates of the world. But only five varieties of these ferns are known to be edible and safe. However, only the young leaves, called fiddleheads, are eaten up to a certain height. The fern species Diplazium esculentum is widespread throughout Asia and Oceania. In Assam, it is available in the market, where it is called Dhekia-khaak, and eaten as a delicacy by all classes of society. This vegetable fern contains nutrients such as iron, calcium, phosphorus, omega 3 and omega 6. This fern (Diplazium esculentum) has a large economic potential worldwide.

Legumes such as peas, chickpeas and dried beans can be prepared in many different ways as meat substitutes. Seitan or wheat gluten is made from wheat and in many ways is regarded as meat of cereal origin. A large number of meat substitutes are solely produced from the legume soybean. There is no limit to meat imitation, and theoretically as well as practically almost all plant foods can be prepared as a meat substitute.

An overview of the food habits of faith communities like the Jains, Buddhists and Hindus provide some new revelations on the subject of the meatless diet. These food cultures are thousands of years old and have a very broad knowledge of how to select and prepare tasty, healthy and varied dishes. The traditional culinary art from antiquity and the art of cooking of world's poorer populations, for whom meat is too expensive, are helpful when trying to give up eating meat. However, the most important element is an awareness of what constitutes a healthy diet.

Some significant steps

Animal husbandry is a braided belt economy which is not easy to abandon overnight. It is a tradition and socio-economic dependency. But the symbiosis between human and domestic animals is long gone. If humans would not enslave animals in such a way, they would live longer, supply milk, eggs and workpower and at the end of their lives, leave their bodies as food in the form of meat. In contrast, in industrial animal farming, animals have a short life during which they have to make a large profit for modern man. Modern animal breeding is unethical, unnatural and harmful to health and nature, and it should be stopped as soon as possible.

The cultivation of fodder and the production of feed are relatively recent activities. The dependency on man for food forces the animals to develop unnaturally. For example, a cow weighing 400 kg produces 30 liters of milk a day, more than the body weight of a newborn calf. Fodder cultivation is the most unnatural form of agriculture. This act is the starting point of meat production and other environmental disasters. An end to the cultivation of fodder crops would mean an end to modern animal husbandry. But it does not mean that the domestic animals would die out. They would still exist, but be allowed to live longer, natural lives, which would be much better than the way they live at present.

Modern man is supplied with food in much the same way as farm animals. More than 96% of the population in industrialized societies have no relationship with agriculture. This tendency to be removed from food production is a global trend. In a few decades, only a small labor force will be active in agriculture, responsible for supplying both man and animal. The animals are processed into meat before diseases caused by over-feeding can break out, but human beings have a normal life-span plagued with diseases and remedies.

Once everybody was involved in the production of food, as an active or passive participant. Even the king hunted and helped with plowing and growing crops. The Emperor of Japan still goes once a year to plant rice. This symbolic act is carried out in annual ceremonies by many nobles in the Far East. Modern food production needs to be re-thought, re-structured and changed. Agriculture must be decentralized completely, and carried out locally and regionally. Long distance transport of food should be used only for low water-containing products. The prime efforts of humanity should be devoted to providing clean drinking water and healthy food for the population. These two undertakings should engage far more people than any other field of industry, trade or services. Food is more important than medical care. Expenditure on defense would gradually be reduced because countries would not be as insecure as half a century before. International borders would be opened, making defense superfluous. We may do away with borders between countries but distancing ourselves further from food production not only damages health but may destroy nature and the future of our offspring.

Bibliography & References

American Journal of Clinical Nutrition; Diet, vegetarianism, and cataract risk, Bethesda, March 2011

Animal Feed Resources Information System/FAO, Rome 2012

BBC World Service: The cost of obesity, 24.03.2014

Becker, J; Hungry Ghosts, London 1996

BIDMC; Meat Choices Linked to Heart Disease and Cancer Deaths, Boston 2009

British Journal of Cancer; Meat consumption and risk of breast cancer, London 2007

Brown, D/Telegraph; The recipe for disaster that killed 80 and left a 5 billion Pound bill, London 2001

Brown, R.L and Woods, R.G; Future Dimensions of World Food and Population, Boulder 1981

Brunnvalla/Sweden - Land und Leute/Wölfe, 2012

Bundesministeriums der Justiz; Tierschutzgesetz, Berlin 2006

Camporesi, P; Bread of Dreams, Chicago 1996

CDC; Fact Sheet, Dog Bite, Atlanta, USA 2013

CDC; Motion Sickness, Atlanta 2015

CIA; World Factbook/Denmark, Natural hazards, Washington, DC, 2013

Culinary Institue of America, New York, 01.03.2011

DEFRA/Gov.UK; Animal and plant health, London 2015

DEFRA/Gov.UK; Animal Health and Welfare, FDM Data Archive, London 2004

Department of Labour; Common Illnesses in the Meat Industry/PDF, Auckland 2013

DGSM; The sleep of women, Berlin, 2011

EFSA; Zoonotic diseases, Parma 2015

EPIC; Key findings – Key results, current & near future scientific activity, Lyon 2007

European Association for Grain Legume Research; Grain Legumes, Paris 2007

European Commission; Control and monitoring programme for Classical Swine Fever, Brussel 2012

European Union bans Brazilian pork supply, agrarheute.com, 26.03.2010

FAA; Sleep and Psychomotor, Performance during Commercial Ultra Long Range Flights, Washington, DC 2008

FAO/AGPC; Grassland and Pasture, Crop Group, Rome 2005

FAO; Animal Production and Health, Rome 2008

FAOSTAT; Production, Livestock & Crops, Rome 2015

FAOSTAT; World Production, Trade and Consumption, Rome 2015

Feedipedia.org, Manure and poultry litter, Paris 2015

Federal Institute for Risk Assessment, No. 016, Berlin, 2009

FDA; Bad Bug BooK/PDF, Silver Spring, 2015

FDA; Bovine Spongiform Encephalopathy, Silver Spring, 2012

Friedrich Loeffler Institut; Tierseuchengeschehen, Greifswald 2015

Gait, Edward; A History of Assam, Calcutta, 1933

German.China.Org.Cn 15th and 03, 2011, lean meat products

GIEWS; Food Outlook – all issues, Rome 2015

Grigg, D; The World Food Problem 1950-1980, Oxford 1985

Grigg, D.B; Population Growth and Agrarian Change, Cambridge 1980

Hägg, G; Presentation Speech, The Nobel Prize in Chemistry 1964

Harris, M; Kannibale und Könige - Wachstumsgrenzen der Hochkulturen, München 1995

Harris, M; Wohlgeschmack und Widerwillen, Stuttgart 1988

Holmgren, I; Presentation Speech, The Nobel Prize in Medicine 1934

International Egg Commission; Egg Industry, London 2015

IUCN; Table 7: Species changing IUCN Red List Status 2013.1, Cambridge 2014

Kaufmann, O; Aus Indiens Dschungel, Erlebnisse und Forschungen, Leipzig 1923

Kent, G; Political Economy of Hunger, New York 1984

Knyvett, H; The Defence of the Realm, Oxford 1906

Kollath, W; Textbook of Hygiene, Vol II, Stuttgart 1949

Livi-Bacci, M; Population and Nutrition, Cambridge 1991

Mallory, W.H; China - Land of Famine, New York 1926

Malthus, TR; An Essay on the Principles of Population, London 1888

Mills, M; The Comparative Anatomy of Eating - Lectures, New York 2006

Milner-Guland, E.J. & Benett, E.L/Trends in Ecology and Evolution; Wild

Meat - The bigger Picture, Voll. 18, No. 7, London 2003

Mountain Voices Org; Livestock in the southwest collection, Online 2007

National Cancer Institute; Chemicals in Meat Cooked at High Temperatures and Cancer Risk, Rockville 2010

National Institutes of Health; Foodborne Illnesses, Bethesda 2014

National Pork Board; Pork Producers Optimistic, Des Moines, Iowa 2015

National Pork Producers Council; International Trade, Washington, DC 2015

Rahn, R/Ball, B; Local Anesthesia in Dentistry, Frankfurt M. 2001

Reuters; Kidney racket scandal in Gurgaon shocks India, New Delhi, Jan 28, 2008 3:26 IST

RIA Novosti, Moscow, 23 11 2012

Statistisches Bundesamt; Zahlen & Fakten, Wiesbaden 2015

Strayer, J.R (Hrg.); Dictionary of the Middle Age, vol. 5, New York 1985

Swift, J; A Modest Proposal, London 1729

Täufel/Ternes/Tunger/Zobel; Lebensmittel-Lexikon, Hamburg 1998

The Book of Leviticus

The American Egg Board; 2013 Annual Report, Park Ridge 2015

The Economist; Paradise well and truly lost, London Dec 20th 2001

The world's biggest diabetes epidemic, China org.cn, 29.08.2011

Trends in Ecology and Evolution, Voll. 18, No. 7, London 2003

UNCCD; Key topics, Bonn 2015

UNCTAD; Handbook of Statistics 2014, Online Version, Geneva 2015

UNEP; Disasters and Conflicts, Geneva 2015

United Nations Framework Convention on Climate Change, GHG data, Bonn 2012

Universität Münster/Haus der Niederlande - Rückblick: Tierseuchen im Grenzgebiet, Münster 2006

UNO; Demographic Yearbook New York 2015

USDA; Disaster and Drought Information, Washington DC 2015

USDA; Feed Grains Database, Washington DC 2015

Whale.to/a/comp.Mills, MR, The Comparative Anatomy of Eating, 07.06.2012

WHO; Dementia, Fact sheet N 362, Geneva, March 2015

WHO, Zoonoses and the Human-Animal-Ecosystems Interface, Geneva, March 2015

WHO, Zoonoses and veterinary public health, Geneva, June 2012

WHO; Water Sanitation Health, Geneva 2014

WTO; U.S. risks trade sanctions in WTO meat label dispute, Geneva 2014

Woodham-Smith, C; The Great Hunger, London 1964

WWF; Protecting Species, Gland 2015

WWF; The wolf hunt in Sweden 2010 and 2011, Stockholm / Solna 2012

Zeitlin, M.F; Nutrition and Population Growth, Cambridge 1982

Zentrum-der-Gesundheit, CH-8008 Zürich 2013

Zischka, A; Brot für 2 Milliarden Menschen, Leipzig 1938

www.ingramcontent.com/pod-product-compliance
Lightning Source LLC
Chambersburg PA
CBHW050056230526
45470CB00004B/1559